T0320421

Generalized Kernel Equating with Applications in R

Generalized Kernel Equating is a comprehensive guide for statisticians, psychometricians, and educational researchers aiming to master test score equating. This book introduces the Generalized Kernel Equating (GKE) framework, providing the necessary tools and methodologies for accurate and fair score comparisons.

The book presents test score equating as a statistical problem and covers all commonly used data collection designs. It details the five steps of the GKE framework: presmoothing, estimating score probabilities, continuization, equating transformation, and evaluating the equating transformation. Various presmoothing strategies are explored, including log-linear models, item response theory models, beta4 models, and discrete kernel estimators. The estimation of score probabilities when using IRT models is described and Gaussian kernel continuization is extended to other kernels such as uniform, logistic, Epanechnikov and adaptive kernels. Several bandwidth selection methods are described. The kernel equating transformation and variants of it are defined, and both equating-specific and statistical measures for evaluating equating transformations are included. Real data examples, guiding readers through the GKE steps with detailed R code, and explanations are provided. Readers are equipped with advanced knowledge and practical skills for implementing test score equating methods.

Marie Wiberg is a professor in statistics with a specialty in psychometrics at Umeå University in Sweden. She is the author of more than 60 peer-reviewed research papers and has edited nine books. Her research interests include test equating, large-scale assessments, parametric and nonparametric item response theory, and educational measurement and psychometrics in general.

Jorge González is an associate professor at the Faculty of Mathematics, Pontificia Universidad Católica de Chile. He is author of a book and several publications on test equating. His research is focused on statistical modeling of data arising from the social sciences, particularly on the fields of test theory, educational measurement, and psychometrics.

Alina A. von Davier is the chief of assessment at Duolingo, and the Founder of EdAstra Tech. She has received several awards, including the ATP's Career Award, the AERA award for significant contribution to educational measurement and research methodology, and the NCME annual award for scientific contributions. Her research is in the field of computational psychometrics, machine learning, assessment, and education.

Chapman & Hall/CRC
Statistics in the Social and Behavioral Sciences Series

Series Editors
Jeff Gill, Steven Heeringa, Wim J. van der Linden, and
Tom Snijders

Recently Published Titles

Mixed-Mode Surveys: Design and Analysis
Jan van den Brakel, Bart Buelens, Madelon Cremers, Annemieke Luiten,
Vivian Meertens, Barry Schouten, and Rachel Vis-Visschers

Applied Regularization Methods for the Social Sciences
Holmes Finch

An Introduction to the Rasch Model with Examples in R
Rudolf Debelak, Carolin Stobl, and Matthew D. Zeigenfuse

Regression Analysis in R: A Comprehensive View for the Social Sciences
Jocelyn H. Bolin

Intensive Longitudinal Analysis of Human Processes
Kathleen M. Gates, Sy-Min Chow, and Peter C. M. Molenaar

**Applied Regression Modeling: Bayesian and Frequentist
Analysis of Categorical and Limited Response Variables
with R and Stan**
Jun Xu

**The Psychometrics of Standard Setting: Connecting Policy and
Test Scores**
Mark Reckase

Crime Mapping and Spatial Data Analysis using R
Juanjo Medina and Reka Solymosi

Computational Aspects of Psychometric Methods: With R
Patricia Martinková and Adéla Hladká

Principles of Psychological Assessment
With Applied Examples in R
Isaac T. Petersen

Multilevel Modeling Using R, Third Edition
W. Holmes Finch, Jocelyn E. Bolin, and Ken Kelley

Polling, Prediction, and Testing, Second Edition
Ole J. Forsberg

Generalized Kernel Equating with Applications in R
Marie Wiberg, Jorge González and Alina A. von Davier

For more information about this series, please visit: https://www.routledge.com/
Chapman--HallCRC-Statistics-in-the-Social-and-Behavioral-Sciences/book-series/
CHSTSOBESCI

Generalized Kernel Equating with Applications in R

Marie Wiberg, Jorge González, and
Alina A. von Davier

CRC Press
Taylor & Francis Group
Boca Raton London New York

CRC Press is an imprint of the
Taylor & Francis Group, an **informa** business

A CHAPMAN & HALL BOOK

Designed cover image: © Marie Wiberg, Jorge González, and Alina A. von Davier

First edition published 2025
by CRC Press
2385 NW Executive Center Drive, Suite 320, Boca Raton FL 33431

and by CRC Press
4 Park Square, Milton Park, Abingdon, Oxon, OX14 4RN

CRC Press is an imprint of Taylor & Francis Group, LLC

© 2025 Taylor & Francis Group, LLC

ISBN: 978-1-138-19698-8 (hbk)
ISBN: 978-1-032-90495-5 (pbk)
ISBN: 978-1-315-28377-7 (ebk)

DOI: 10.1201/9781315283777

Typeset in CMR10
by KnowledgeWorks Global Ltd.

Publisher's note: This book has been prepared from camera-ready copy provided by the authors.

To Hanna and Emelie
- Marie Wiberg

To Fanny, Javiera, and Renato
- Jorge González

To Paul Holland
- Alina von Davier

Contents

Foreword   xiii

Preface   xv

Acronyms   xix

Symbols   xxiii

I Test Equating and Kernel Equating Overview   **1**

1 Introduction   **3**
   1.1 Test Score Equating 4
   1.2 Test Equating as a Statistical Modeling Problem 6
      1.2.1 Notation 6
      1.2.2 Equating as a Statistical Model 7
   1.3 The Target Population and Data Collection Designs 9
      1.3.1 The Target Population 9
      1.3.2 Data Collection Designs 10
      1.3.3 Equivalent Groups Design 10
      1.3.4 Single Group Design 10
      1.3.5 Counterbalanced Design 11
      1.3.6 Non-Equivalent Groups with Anchor Test Design . . . 12
      1.3.7 Non-Equivalent Groups with Covariates Design 12
   1.4 Equating Transformations 13
      1.4.1 The Kernel Equating Function 14
   1.5 Evaluating the Equating Transformation 15
   1.6 **R** Packages Used in This Book 15
   1.7 Empirical Examples 16
   1.8 Summary and Overview of the Book 16

2 Kernel Equating   **19**
   2.1 Introduction 19
      2.1.1 The Five Steps in KE 21
   2.2 The Generalized Kernel Equating Framework 22
      2.2.1 Presmoothing 23
      2.2.2 Estimating Score Probabilities 25

        2.2.3   Continuization 26
                  2.2.3.1   Bandwidth Selection 26
        2.2.4   The Equating Transformation 27
        2.2.5   Evaluating the Equating Transformation 27
  2.3 Summary 29

II   Generalized Kernel Equating Framework   31

3 Presmoothing   33
  3.1 Presmoothing the Data 33
  3.2 Parametric Statistical Models for Presmoothing 35
        3.2.1   Polynomial Log-Linear Models 36
  3.3 Nonparametric Methods for Presmoothing 38
        3.3.1   Nonparametric Discrete Kernel Estimators 39
  3.4 Smoothing using Mixture Distributions 40
        3.4.1   Beta4 Models 41
        3.4.2   Item Response Theory Models 42
                  3.4.2.1   Binary Item Response Theory Models 43
                  3.4.2.2   Polytomous Item Response Theory Models . 44
  3.5 Evaluation of Presmoothing Methods 45
        3.5.1   Log-Linear Models 46
        3.5.2   Discrete Kernel Estimators 47
        3.5.3   Beta4 Models 48
        3.5.4   IRT Models 48
        3.5.5   Choosing a Presmoothing Model 52
  3.6 Summary 52

4 Estimating Score Probabilities   55
  4.1 Design Functions 56
        4.1.1   EG Design 57
        4.1.2   SG Design 58
        4.1.3   CB Design 59
        4.1.4   NEAT Design 59
        4.1.5   NEC Design 62
        4.1.6   Comparison of Designs 64
  4.2 Estimated Probabilities 64
  4.3 Estimated Probabilities from IRT Models 66
        4.3.1   The Lord-Wingersky (LW) Algorithm 66
        4.3.2   The Poisson's Binomial Distribution 67
        4.3.3   Other Approximate and Exact Methods 68
        4.3.4   The LW Algorithm for Polytomous Data 70
        4.3.5   Conditional Probabilities 70
  4.4 Summary 71

5 Continuization **73**
  5.1 Kernel Density and Distribution Estimation 74
  5.2 Convolutions 75
  5.3 General Kernel Continuization 77
  5.4 Different Kernels 78
      5.4.1 Gaussian Kernel 79
      5.4.2 Logistic Kernel 79
      5.4.3 Uniform Kernel 80
      5.4.4 Epanechnikov Kernel 81
      5.4.5 Adaptive Kernels 81
      5.4.6 The Percentile Rank Method 83
  5.5 Cumulants 84
  5.6 Summary 86

6 Bandwidth Selection **87**
  6.1 Bandwidth Selection in Kernel Density Estimation 88
  6.2 Selecting Bandwidths 88
  6.3 Minimizing a Penalty Function 90
  6.4 Leave-One-Out Cross-Validation 91
  6.5 Likelihood Cross-Validation 92
  6.6 Double Smoothing 93
  6.7 Rule-Based Bandwidth Selection 94
  6.8 A CDF-Based Bandwidth Selection Method 95
  6.9 Summary 96

7 Equating **97**
  7.1 Assumptions 98
  7.2 Score Comparability through Scale Transformation 100
  7.3 Equating Transformations for the OSE Framework 101
  7.4 Linear Equating Transformation 103
  7.5 Equating Transformation According to Equating Design . . . 104
      7.5.1 Chained Equating 105
  7.6 IRT Calibration 107
      7.6.1 Concurrent Calibration 109
  7.7 Equating Transformations for the Local Equating Framework 110
  7.8 Graphical Representation of the Equating Transformation . . 111
  7.9 Summary 111

8 Evaluating the Equating Transformation **113**
  8.1 Equating-Specific Measures 113
      8.1.1 Difference That Matters 114
      8.1.2 Percent Relative Error 114
      8.1.3 First-Order and Second-Order Equity 114
      8.1.4 Standard Error of Equating 115
          8.1.4.1 The Delta Method to Obtain the SEE 116

        8.1.4.2   Bahadur Representation to Obtain the SEE   122

   8.1.5   Bootstrap Standard Error of Equating   124

   8.1.6   Indices to Compare Equating Transformations   125

        8.1.6.1   Mean Signed Difference   126

        8.1.6.2   Mean Absolute Difference   126

        8.1.6.3   Root Mean Squared Difference   126

        8.1.6.4   Standard Error of Equating Difference   126

8.2   Statistical Measures   127

   8.2.1   Bias   128

   8.2.2   Mean Squared Error   128

   8.2.3   Root Mean Squared Error   129

   8.2.4   Standard Error   129

   8.2.5   Cumulants   130

8.3   Simulating Test Scores   130

   8.3.1   Simulated Equating-Specific Measures   132

   8.3.2   Simulated Comparison Indices   134

   8.3.3   Examples of Statistical and Simulated Measures . . .   135

8.4   Choice of Evaluation Measure   136

8.5   Summary   137

III Applications   139

9 Examples under the EG design   141

9.1   Software Choice   141

9.2   SEPA data   142

   9.2.1   Preparing the SEPA Data for **SNSequate**   142

9.3   Step 1: Presmoothing   143

   9.3.1   Beta4 Models   144

   9.3.2   Discrete Kernel Estimators   145

9.4   Step 2: Estimating Score Probabilities   147

   9.4.1   Beta4 Models   147

   9.4.2   Discrete Kernel Estimators   148

9.5   Step 3: Continuization   150

   9.5.1   Bandwidth Selection   150

   9.5.2   Kernel Selection   152

9.6   Step 4: Equating   152

9.7   Step 5: Evaluating the Equating Transformation   156

   9.7.1   PRE   156

   9.7.2   Bootstrap Standard Error of Equating   158

   9.7.3   Freeman-Tukey Residuals   160

9.8   Summary   162

10 Examples under the NEAT design **163**
10.1 Software Choice 163
10.2 ADM Data 164
    10.2.1 Preparing the ADM Data for **kequate** 165
10.3 Simulated Polytomous Data 166
10.4 Step 1: Presmoothing 168
    10.4.1 Log-Linear Models 168
        10.4.1.1 Modeling Complexities 170
        10.4.1.2 Log-Linear Model Fit 171
    10.4.2 Binary IRT Models 173
        10.4.2.1 IRT Model Fit 175
    10.4.3 Polytomous IRT Models 181
        10.4.3.1 IRT Model Fit 182
10.5 Step 2: Estimating Score Probabilities 185
    10.5.1 Log-Linear Models 185
    10.5.2 IRT Models 186
10.6 Step 3: Continuization 188
    10.6.1 Bandwidth Selection and Kernel Selection 188
10.7 Step 4: Equating 188
    10.7.1 Log-Linear Presmoothed Data 190
    10.7.2 IRT Presmoothed Data 194
10.8 Step 5: Evaluating the Equating Transformation 198
    10.8.1 PRE 198
    10.8.2 SEE 199
    10.8.3 Bootstrap Standard Error of Equating 199
    10.8.4 SEED 201
    10.8.5 MSD, MAD, and RMSD 203
10.9 Summary 204

IV   Appendix **205**

A Installing R and Reading in Data **207**
A.1 Installing **R** for the First Time 207
    A.1.1 **Rstudio** 207
A.2 Installing and Using **R** packages 208
A.3 Loading Data 208

B R packages for GKE **211**
B.1 Presmoothing 211
B.2 Estimating Score Probabilities 212
B.3 Continuization 212
B.4 Bandwidth Selection 212
B.5 Equating 213
B.6 Evaluating the Equating Transformation 213

Bibliography   **215**

Index   **235**

Foreword

In the dynamic area of educational measurement today, the kernel method of test equating, pioneered by von Davier, Holland, and Thayer two decades ago in 2004, remains a transformative publication. Their groundbreaking work has not merely altered the landscape of equating methodologies; it has introduced a new era of precision, flexibility, and relevance in the pursuit of fair and accurate comparisons across diverse test forms. Come 2024, the comprehensive exploration undertaken by Wiberg, González, and von Davier in this volume (2024) extends the legacy of the kernel method, propelling it into a generalized framework to meet the evolving challenges of educational assessment.

The kernel method, rooted in the principles of kernel smoothing borrowed from nonparametric statistics, diverges from traditional equating paradigms by offering a more principled representation of true score distributions. By embracing this departure from classical models, Wiberg, González, and von Davier guide the readers into a realm where the rigidity of assumptions is replaced by a more robust and flexible solution. The heart of the method lies in its adept use of kernels to smooth observed score distributions, effectively mitigating the impact of measurement error and sampling variability.

This book serves as an invaluable guide, navigating readers through the mathematical underpinnings and practical implementations of the generalized kernel equating method. From theoretical foundations to real-world applications, Wiberg, González, and von Davier illuminate the path for researchers, practitioners, and educators seeking a profound understanding of test equating in the contemporary educational assessment landscape.

What distinguishes this work is the integration of a wide variety of equating methods and designs in what is the comprehensive framework of generalized kernel equating. As we go through the chapters, it becomes clear that this method not only enhances equating accuracy but also accommodates the complexities inherent in modern testing scenarios. The authors' exploration of real-world applications, software availability, and empirical studies attests to the versatility and efficacy of the generalized kernel method across diverse testing contexts, cementing its relevance for educational measurement.

In conclusion, Wiberg, González, and von Davier's contribution goes beyond the theoretical, by providing practitioners with a powerful tool to navigate the challenges of equating in an era where educational assessments play an increasingly pivotal role in shaping individual trajectories and societal outcomes. This book opens the path towards a more nuanced understanding of test equating through the lens of the generalized kernel method. As you, the reader, embark on this journey, you will experience the authors' profound impact on our theoretical foundations and their legacy in the ever-evolving landscape of educational measurement.

Gunter Maris,
Westervoort, Netherlands, June 2024

Preface

In 2004, von Davier, Holland, and Thayer published a methodological book on equating test scores using a Gaussian kernel. That book has been extremely influential on the theory and practice of test equating. In 2011, von Davier edited a volume on statistical models for test equating comprising of some of the research inspired by the von Davier et al. (2004) book. A few years ago, González and Wiberg (2017) published an applied book on test equating in which it is shown how both traditional and more modern methods such as kernel equating (KE) can be implemented and applied in practice using the **R** software (R Core Team, 2024). From these books, it is evident that the use of kernels, as well as modern statistical models and methods for test equating, could be beneficial for the accuracy of test equating results. This year marks the 20th anniversary of the first KE book, and in this book, we aim to highlight the developments that have occurred since then and are still ongoing.

Building on our previous experiences, this book extends the (Gaussian) KE methodology and makes the sophisticated theoretical models accessible to practitioners and researchers by providing the **R** code (R Core Team, 2024) and packages necessary to implement them. We provide an up-to-date overview of the KE process and methodologies and synthesize recent work on KE into an organized and comprehensive framework that we call the *generalized kernel equating* (GKE) framework.

In this book, we expand the Gaussian KE from von Davier et al. (2004) in multiple directions, from incorporating new methodologies at each of the equating steps, to providing a unified framework for these expansions that may support new and specific applications or research inquiries. We believe that this coherent and integrated framework that explicitly connects the assumptions, different and separated equating methods, and steps necessary to implement them will help researchers and practitioners avoid model misspecification or overfitting.

Similarly to our previous work, under the GKE framework we continue to situate the test equating process as a statistical modeling process. This includes test and sampling design, measurement and modeling issues, assumptions and requirements made at different stages of the process, as well as the equating methods under the selected equating model. Equating models that use a total test score as a unit of measurement, such as those included in the observed-score equating (OSE) framework (von Davier, 2013), and models that use the item-person interaction as a unit of measurement, as in item response theory (IRT) models, both use equating transformations with parameters to be estimated and assessed from sample data. In this new book both OSE and IRT equating models are unified under the GKE framework.

The GKE framework extends the theoretical underpinnings of KE and provides an integrated theory of the KE framework by generalizing all its steps. In particular, we propose i) the use of different models and methods for presmoothing, ii) to extend the use of design functions with methods that allow the estimation of score probabilities when different equating models are used, iii) to provide a new perspective on continuization by introducing the concept of convolutions and describing the use of different kernels beyond the Gaussian kernel, iv) to introduce several alternatives to the selection of bandwidth parameters, v) to enable the use of other data besides binary-scored data, and vi) to provide a new perspective on equating evaluation. Using real data, we also provide applications and software code to facilitate the accessibility to and widespread use of the described methods.

This book differs from the volume of González and Wiberg (2017) in that the latter focused on the practical applications of not only KE methods but also several equating methods using **R** code, whereas in this work, we intend to provide a theoretical foundation for GKE. Nevertheless, the applications discussed in this book will draw on González and Wiberg (2017) and on their **R** packages, **kequate** (Andersson et al., 2013), and **SNSequate** (González, 2014).

Intended audience
We believe that multiple groups will find this book of interest. Our intended audience includes researchers, graduate students, test developers, psychometricians, and other professionals with an interest in test equating.

Organization of the book
The book is organized into three parts. The first part, which includes Chapter 1 and Chapter 2, introduces the definitions and the basic elements of the GKE framework.

The second part consists of Chapters 3–8, which describe different steps of the GKE framework. In Chapter 3, we introduce different presmoothing modeling approaches, including parametric statistical models, nonparametric methods, and methods based on mixture distributions. Chapter 4 illustrates the estimation of the test score probabilities, expanding the use of design functions with methods suitable for IRT equating models. In Chapter 5, the continuization step is thoroughly re-examined using the theory of convolutions, allowing for the use of different kernels beyond the Gaussian. In Chapter 6, different methods for selecting the bandwidth (the weight of the kernel) are described. Chapter 7 is focused on the equating transformations calculated using either marginal or conditional score distributions, and also calculated as a functional composition of various equating transformations. Chapter 8 is focused on the evaluation of an equating transformation both through equating-specific measures and through simulations and statistical measures.

The third and last part of the book is composed of two chapters focused on applications using real data sets for illustrations. The empirical examples and **R** code are provided for the equivalent groups design in Chapter 9, and for the nonequivalent groups with anchor test design in Chapter 10. Finally, Appendix A contains a brief description of how to install **R** packages and read in test data, and Appendix B contains short descriptions of which **R** packages can be used to implement the different steps in the GKE process.

Acknowledgements

We want to acknowledge the importance of real-life creative meetings. The idea of this book emerged when the three of us met in 2015 during the ISI World Statistics Congress and had lunch at a small restaurant at the Botanical Garden in Rio de Janeiro, Brazil. Although most of our book meetings since then have been digital, important creative meetings have taken place during our writing process on all three continents on which we live. Thus we also would like to acknowledge the Swedish Research Council (grant 2018-03995) for funding some of these crucial physical meetings.

Marie Wiberg	Umeå, Sweden,	June 2024
Jorge González	Santiago, Chile,	June 2024
Alina A. von Davier	Boston, USA,	June 2024

Acronyms

AIC: Akaike Information Criterion

AMAD: Average Mean Absolute Difference

AMISE: Asymptotic Mean Integrated Squared Error

AMSD: Average Mean Signed Difference

APAD: Average Point Absolute Difference

APSD: Average Point Signed Difference

APRSD: Average Point Root Squared Difference

ARMSD: Average Root Mean Squared Difference

BIC: Bayesian Information Criterion

CB: Counterbalanced

CDF: Cumulative Distribution Function

CE: Chained Equating

CV: Cross-Validation

DF: Design Function

DS: Double Smoothing

DTM: Difference That Matters

ECDF: Empirical Cumulative Distribution Function

EDIFF: Equating Difference

EPDF: Empirical Probability Density Function

EG: Equivalent Groups

FOE: First-Order Equity

GKE: Generalized Kernel Equating

GPCM: Generalized Partial Credit Model

GRM: Graded Response Model

Hcrit: Haebara criterion

ICC: Item Characteristic Curve

IRT: Item Response Theory

ISE: Integrated Squared Error

KE: Kernel Equating

LCV: Leave-one-out Cross-Validation

LiCV: Likelihood Cross-Validation

LW: Lord-Wingersky

MAD: Mean Absolute Difference

MGF: Moment Generating Function

MISE: Mean Integrated Squared Error

MLCV: Maximum Likelihood Cross-Validation

MSD: Mean Signed Difference

MSE: Mean Squared Error

NEAT: Non-equivalent groups with Anchor Test

NEC: Non-equivalent groups with Covariates

OSE: Observed-Score Equating

PCM: Partial Credit Model

PDF: Probability Density Function

PLCV: Penalty Leave-one-out Cross-Validation

PS: Propensity Score

PMF: Probability Mass Function

PRE: Percent Relative Error

PRM: Percentile Rank Method

PSE: Post-Stratification Equating

QB-SEE: Quantile-based Standard Error of Equating

RMSD: Root Mean Squared Difference

RMSE: Root Mean Squared Error

RMSEA: Root Mean Squared Error of Approximation

SOE: Second-Order Equity

SE: Standard Error

SEE: Standard Error of Equating

SEED: Standard Error of Equating Difference

SG: Single Group

SLcrit: Stocking-Lord criterion

SRT: Silverman's Rule of Thumb

1PL: One-parameter logistic IRT model

2PL: Two-parameter logistic IRT model

3PL: Three-parameter logistic IRT model

Symbols

Listed below are symbols used together with a short description and the page number when the symbol first appears in the book. There are a few symbols that have multiple uses, but care was taken to avoid any confusion in the text.

Symbol	Description	Page
$\mathbf{1}_J$	J dimensional column vector of ones	58
$\mathbf{1}_K$	K dimensional column vector of ones	58
$\mathbf{1}_{S_x}$	Indicator function used in discrete kernels	39
α	Sensitivity parameter used in adaptive kernels	82
α	Shape parameter in the Beta distribution	131
β	Shape parameter in the Beta distribution	131
β_0	Parameter acting as a normalization constant in a log-linear model	36
δ	The Dirac delta function	75
Δ	Matrix used to define design functions which can differ between different designs	56
$\partial\varphi/\partial\mathbf{r}$	The derivatives of the equating transformation with respect to $\mathbf{r}$	117
$\partial\varphi/\partial\mathbf{s}$	The derivatives of the equating transformation with respect to $\mathbf{s}$	117
ϵ	The value of the interval limits where a uniform random variable is defined	80
η	Index to denote the ηth cumulant	85
γ_j	Item pseudo guessing parameter for item j in IRT models	44
κ	Constant in the bandwidth selection penalty function	91
Φ	CDF of the standard normal distribution	68
ϕ	Density of the standard normal distribution	69
φ	The equating transformation	7
λ_j	Local bandwidth factor used in adaptive kernels	82
λ	Parameter indexing the family of local equating transformations	70
Λ	Set of possible values for λ	70
μ_X, μ_Y	Mean of X and Y, respectively	13
ν	Treatment variable, indicating which test form (X or Y) is administered	63

Symbol	Description	Page
ω_j	Parameter vector of item characteristics for item j	43
Ω	The set of all score probability vectors	56
Π	Elements in $\mathscr{P}$, which are vectors or matrices depending on the design	56
$\psi(\tau)$	Population distribution of proportion-correct true scores	42
$\psi(\theta)$	Conditional expected equating score	115
Ψ	Family of local equating transformations	110
σ_X, σ_Y	Standard deviation of X and Y	13
σ_θ^2	Variance of latent ability	43
$\Sigma_{\hat{R}}$	Estimated covariance matrix of R	118
$\Sigma_{\hat{S}}$	Estimated covariance matrix of S	118
$\Sigma_{\hat{r},\hat{s}}$	Estimated covariance matrix of r and s	116
θ	Latent ability	43
θ	A parameter	7
Θ	Parameter space	7
$\mathcal{A}$	Sample space for anchor score A	12
$\mathcal{B}$	Number of moments in PRE	114
$\mathcal{C}$	The common items set between test forms X and Y	109
$\mathcal{F}$	Family of probability distributions	7
$\mathcal{K}(t)$	Cumulant generating function	84
$\mathscr{L}$	The likelihood ratio chi-square statistic	46
$\mathcal{M}(\theta)$	Model parametrized by θ	24
$\mathbb{N}$	Integer numbers	8
$\mathscr{P}$	The set of all possible population score distributions	56
$\mathcal{P}$	Population P	10
$\mathcal{Q}$	Population Q	10
$\mathbb{R}$	Set of real numbers	13
$\mathcal{T}$	The target population	9
$\mathcal{X}$	Sample space for score X	6
$\mathcal{Y}$	Sample space for score Y	6
$\mathcal{Z}$	Sample space for score Z	105
a	Anchor test score	12
a_j	Item discrimination parameter for item j in IRT models	44
a	Integer arm in the triangular discrete kernel estimator	39
A	Anchor test form	12
A	Random variable for anchor test score	12
A	Equating coefficient in IRT parameter linking	108
A_j	Indicator variable in $\text{PEN}_2(h_X)$	90
b_j	Item difficulty for item j in IRT models	44
$b_{j,v}$	Step difficulty parameters in the GPCM	45
b_{ij}	Constants in the log-linear models	36

Symbol	Description	Page
b	Vector of constants used to specify a log-linear model for score probabilities	36
$\mathbf{B}$	Matrix of constants used to specify a log-linear model for score probabilities	36
B	Equating coefficient in IRT parameter linking	108
$B_{x,h}$	The discrete kernel associated with the binomial distribution	39
c	Categories when using the Dirac discrete kernel estimator	40
C	Covariates	12
$\mathbf{C}_R$	The C-matrix factor of Σ_R	119
$\mathbf{C}_S$	The C-matrix factor of Σ_S	119
d	Index to denote a stratified propensity score	63
D	Number of strata used in the NEC design	121
D	Scaling constant in IRT models	44
$\mathbf{D}_{\sqrt{r}}$	Diagonal matrix with vector $\sqrt{r}$ along the main diagonal.	119
$\mathbf{D}$	Vector of covariates	63
df	Degrees of freedom	51
$\mathbf{DF}(\cdot)$	Design function	25
$D_{x,h,c}$	The discrete kernel associated with the Dirac discrete uniform random variable	40
$e_{Xd},\ e_{Yd}$	Stratified propensity score d of test forms X and Y, respectively	121
E_n	Random variable for the total number of successes in n independent Bernoulli trials	67
$\mathrm{E}(X_{ij})$	Expected value of X_{ij}	50
$\hat{f}_{g_X}(x)$	Estimate of the score density with a pilot bandwidth g_X	93
$f_{h_X}(x)$	Density function of the continuized random variable $X(h_X)$	20
$\hat{f}_{h_X}^{(-j)}(x_j)$	Estimate of the density function based on the subsample when $(x_j, \hat{r}_j)$ have been left out	91
$f_{h_Y}(y)$	Density function of the continuized random variable $Y(h_Y)$	20
$f_l(x \mid \theta)$	The conditional distribution of sum scores over the first l items for test takers of ability θ	66
$f_N(x, h)$	Estimator of $f(x)$	47
$f(\theta)$	Distribution of the latent ability	43
$\hat{f}_{N,h,B}(x)$	The binomial kernel estimator	39
$\hat{f}_{N,h,D}(x)$	The Dirac discrete uniform kernel estimator	40
$\hat{f}_{N,h,T}(x)$	The discrete triangular kernel estimator	39
F_{AP}	Anchor test score distribution in population $\mathcal{P}$	106

Symbol	Description	Page
F_{AQ}	Anchor test score distribution in population $\mathcal{Q}$	106
$F_{h_X}(x)$	CDF of the continuized random variable $X(h_X)$	14
$F'_{h_X}(x)$	The derivative of $F_{h_X}(x)$, which is equal to the density f_{h_X}	117
$F_{h_Y}(y)$	CDF of the continuized random variable $Y(h_Y)$	14
F_X	The CDF of X	14
$F_X(x)$	The CDF of X	6
$F_{X|\theta}$	Conditional distribution of X given latent ability θ	71
F_Y	The CDF of Y	14
$F_Y(y)$	The CDF of Y	6
F'_Y	The derivative of F_Y, i.e., the density f_Y	117
$F_{Y|\theta}$	Conditional distribution of Y given latent ability θ	71
FT	Freeman-Tukey residual	46
g	Index used in the GPCM	45
$\bar{g}$	Geometric mean in the definition of a local bandwidth factor	82
G	Group from the target population	10
G_1, G_2	Group 1 and Group 2 from the target populations	10
h_X, h_Y	The bandwidths that are used for continuization of $F(x)$ and $F(y)$, respectively.	14
$\mathbf{i}$	The imaginary unit	69
i	Index for test taker i	11
I	Combinations of covariates values in the NEC design	62
$\mathbf{I}_J$	$J \times J$ identity matrix	57
$\mathbf{I}_K$	$K \times K$ identity matrix	57
j	Index for item j	6
j	Index for score j	6
J	Total number of possible scores in test form X	6
$\mathbf{J}_\varphi$	Jacobian of the equating transformation	116
$\mathbf{J}_{\mathrm{DF}}$	Jacobian of the design function	118
K	Total number of possible scores in test form Y	6
K	Kernel in the GKE framework	78
$K_{x,h}$	Associate discrete kernel	39
L	Total number of possible scores in test form A	12
L	Number of replicates	124
L_r, L_a	Sum limits that serve to accommodate cross moments in polynomial log-linear models	37
m_j	Number of response categories of item j	44
$M_X(t)$	Moment generating function	84
$\mathbf{M}$	A matrix containing zeros and ones used to calculate the column vector of the row sums of a bivariate score probability matrix, Π, from the vectorized version, vec($\mathbf{P}$). Subscripts are used in different designs.	58
$\mathbf{M}_P, \mathbf{M}_Q$	$J \times JL$ and $K \times KL$ matrices	60

Symbol	Description	Page
N	A matrix containing zeros and ones used to calculate the column vector of the row sums of a bivariate score probability matrix, Π, from the vectorized version, $\text{vec}(P)$. Subscripts are used in different designs.	58
N_P, N_Q	$L \times JL$ and $L \times KL$ matrices	60
n	Number of items	6
n_a	Number of common (anchor) items	109
n_x, n_y	Number of items in test form X and Y, respectively	6
N	Sample size of test takers	11
N_j, N_k	Random variable representing number of test takers scoring test score x_j, and y_k respectively	6
N_x, N_y	Sample sizes of test takers administered test forms X and Y, respectively. Subscripts may change to deal with the different designs.	6
p	Number in $[0, 1]$ to represent a proportion	7
p_{ij}	The probability of a test taker i answering a binary-scored item j correctly	44
$p_{ij}(\theta)$	The probability of a test taker i with ability θ answering a binary-scored item j correctly	49
p_j	Probability of scoring x_j	36
p_{jk}	$\Pr(X = x_j, Y = y_k)$	24
p_{jl}	$\Pr(X = x_j, A = a_l)$	24
$p_{(12)jk}$	$\Pr(X_1 = x_j, Y_2 = y_k)$	24
$p_{(21)jk}$	$\Pr(X_2 = x_j, Y_1 = y_k)$	24
$p_{j,x}(\theta)$	Probability of a randomly chosen test taker with ability θ scoring x when there are m_j response categories on item j	44
p	Vector of score probabilities in population $\mathcal{P}$	24
P	Probabilities defining score distributions in the population for each equating design	57
$\text{vec}(P)$	Vectorization of P	58
$\text{PEN}(h_X)$	Penalty function to find h_X.	91
$\text{PEN}_1(h_X)$	First term in the penalty function to find an optimal bandwidth h_X in the continuization process.	90
$\text{PEN}_2(h_X)$	Second term in the penalty function to find an optimal bandwidth h_X in the continuization process.	90
$\Pr(X = x_j)$	Probability of scoring x	36
$\Pr(X = x, A = a)$	The joint probability of $X = x$ and $A = a$	59
$\Pr(Y = y, A = a)$	The joint probability of $Y = y$ and $A = a$	59
$p(\theta_i, \omega_j)$	Item response function with ability θ_i and item parameters ω_j	44
q	Vector of score probabilities in population $\mathcal{Q}$	62

Symbol	Description	Page
Q	Probabilities defining score distributions in the population for each equating design	59
$\text{vec}(Q)$	Vectorization of Q	60
Q_3	Correlation magnitude	177
r_j	Score probabilities for score X	6
r_P	Score probabilities for X scores in populations $\mathcal{P}$	61
r_Q	Score probabilities for X scores in populations $\mathcal{Q}$	61
$r_P(x_j \mid a_l)$	Conditional score probabilities $\Pr(X = x_j \mid A = a_l)$ defined in $\mathcal{P}$	61
$r_Q(x_j \mid a_l)$	Conditional score probabilities $\Pr(X = x_j \mid A = a_l)$ defined in $\mathcal{Q}$	61
r	Vector of score probabilities from test form X	20
r_P	The J-dimensional vector of score probabilities defined in population $\mathcal{P}$	60
R	Presmoothed data that arises in any design	117
$\hat{R}$	Estimated presmoothed data that arises in any design	117
s	Scale parameter in the logistic kernel	80
s_k	Score probabilities for score Y	6
s_P	Score probabilities for Y scores in population $\mathcal{P}$	61
s_Q	Score probabilities for Y scores in population $\mathcal{Q}$	61
$s_P(y_k \mid a_l)$	Conditional score probabilities $\Pr(Y = y_k \mid A = a_l)$ defined in $\mathcal{P}$	61
$s_Q(y_k \mid a_l)$	Conditional score probabilities $\Pr(Y = y_k \mid A = a_l)$ defined in $\mathcal{Q}$	61
s	Vector of score probabilities from test form Y	20
s_Q	The K-dimensional vector of score probabilities defined in population $\mathcal{Q}$	60
S	Presmoothed data that arises in any design	117
$\hat{S}$	Estimated presmoothed data that arises in any design	117
t_l	Score probabilities for A scores	60
t_{Pd}	Score probabilities for different propensity score strata in population $\mathcal{P}$	121
t_{Qd}	Score probabilities for different propensity score strata in population $\mathcal{Q}$	121
t_{Pl}	The anchor score probabilities $\Pr(A = a_l)$ defined in population $\mathcal{P}$	60
t_{Ql}	The anchor score probabilities $\Pr(A = a_l)$ defined in population $\mathcal{Q}$	60
t_P, t_Q	L-dimensional vectors of anchor scores probabilities defined in $\mathcal{P}$ and $\mathcal{Q}$, respectively	60
$T_{a,h,x}$	The discrete kernel associated with the triangular random variable	39

Symbol	Description	Page
T_a	Highest polynomial degree for a_l in a log-linear model	37
T_r	Highest polynomial degree for x_j in a log-linear model	36
u	Number of parameters estimated in log-linear models, used in AIC and BIC	46
v_i	(Un)weighted mean-square for person i	51
v_j	(Un)weighted mean-square for item j	50
V	Random variable representing the frequency for a combination of covariates. Used in the NEC design.	62
V	Continuous random variable	77
$\mathbf{V_P, V_Q}$	Matrices used to calculate the covariances to obtain SEE in the NEC design	122
$\mathrm{Var}(X_{ij})$	Variance of X_{ij}	50
w	In the NEAT design, weight in the synthetic population	60
w_X, w_Y	In the CB design, weight given to the data that are not subject to order effects.	59
W	Continuous random variable	75
W_{ij}	Weight for the item and person infit measures	51
x_j	A possible score value for X	6
x_{jk}	An earned score in category k of item j	70
$\mathbf{x}$	Vector of responses	49
X	Test form X	6
X	Random variable representing test scores from test form X	6
$X(h_X)$	Continuized score random variable of X	77
x_i	The score for test taker i taking test form X	10
$y_{i'}$	The score for test taker i' taking test form Y	10
y_k	A possible score value for Y	6
Y	Test form Y	6
Y	Random variable representing test scores from test form Y	6
$Y(h_Y)$	Continuized score random variable of Y	84
Z	Test form Z	105
Z	Random variable	75
Z_h	Standardized maximum likelihood indices for each person	178

Part I

Test Equating and Kernel Equating Overview

1

Introduction

Test equating methods are statistical procedures employed to account for differences between test forms in standardized testing programs. These test forms are typically constructed based on the same specifications and administered under standardized conditions. The primary objectives of test equating are to ensure that which specific test form test takers take is a matter of indifference, and to maintain the interpretability and comparability of reported test scores. Over the years, numerous research papers and books have been dedicated to the development and exploration of methodologies for test score equating.

The objective of this book is to enhance Gaussian kernel equating (KE), as introduced in von Davier et al. (2004), by expanding it in various directions. This expansion encompasses the incorporation of new methodologies into each of the previously defined KE steps. The aim is to provide a unified framework for these expansions, making it easier for researchers and practitioners to integrate new approaches and address specific applications or research inquiries within the KE context.

This integrated framework serves the purpose of connecting assumptions, different models, and various equating methods, while also offering necessary descriptions on how to implement them in order to assist researchers and practitioners in avoiding model misspecification or overfitting. Chapters 1–2 of the book present definitions and fundamental elements of this integrated framework, while Chapters 3–8 introduce a range of new methods that can be valuable in different situations. The choice of which alternative to use ultimately depends on the specific practical settings and challenges faced in equating applications.

In this introductory chapter, we will begin by providing a brief introduction to the test equating process, drawing from sources such as Kolen and Brennan (2014, Chapter 1) and von Davier (2011). We will also describe equating requirements and challenges that arise in practical applications of equating methodology. A brief review of common data collection designs and their underlying assumptions will be presented. Finally, in the last section, we will offer an overview of the subsequent chapters included in this book.

1.1 Test Score Equating

Test score equating refers to a set of methodologies used primarily in educational assessment, in connection with large, standardized testing programs. However, test equating has started to spread to other areas, such as medicine (Handing et al., 2021), health sciences (Adroher et al., 2019), and neuropsychology (Gross et al., 2019). Methods for test score equating were developed to be able to compare test results from different test occasions, and to help the users of the test results to fairly compare the scores of the test takers who took different test forms. Test score equating is a method rooted in the belief in fair opportunities for all test takers, regardless of test form. There are still educational systems in the world in which high-stakes tests are not standardized or equated, and in which students can take a college admissions test only once a year; if they fail that test, they have to wait another full year before they can take it again. In addition, the number of seats for admission often varies across years. Consequently, the decisions for college admissions are made once a year, based on the results of the "big test" and on the number of seats available. If one is unlucky and is administered a more difficult test form, one may perform worse than one's peers, and have to take the test again a year later. This inequity may lead to validity issues for the test results: are the students who were accepted into college by taking an easy test form similar in ability to those who got in based on a difficult test form? Hence, adjusting for test form differences, i.e., equating test forms, becomes a matter of fairness and validity (González and Gempp, 2021).

The goal of test equating is to treat scores from different test forms as interchangeable and to preserve the meaning of the reported test score. For example, a score of 700 on the $\text{SAT}^{(R)}$ has the same meaning regardless of the form administered or when the test was taken. For test scores to be considered interchangeable, five requirements on the tests to be equated and on the equating function are widely viewed as necessary (Dorans and Holland, 2000; Holland and Dorans, 2006). Some of these requirements originate from Angoff (1971), Lord (1980), and Petersen et al. (1989). However, note that some of them have been criticized by, for instance, van der Linden (2013). Here, we provide a brief summary of the requirements and leave a more thorough discussion of them to Section 7.1 when we have gained more insight into different aspects of equating. The five requirements are:

- The equal construct requirement: The two test forms being equated should measure the same construct (i.e., ability, proficiency, or latent trait).

- The equal reliability requirement: The two test forms being equated should have equal reliability.

- The symmetry requirement: The transformation for mapping the scores from an *origin* test scale to their equivalents on a *target* test scale should be the inverse of the transformation for mapping the scores, reversing the order of the scales.

- The equity requirement: It should be a matter of indifference to a test taker as to which of the test forms he or she answered.

- The population invariance requirement: The equating function should be independent of any subpopulation (differing in some way from the target population) on which it may be computed.

Therefore, the process of equating tests involves not only the adjustment of scales of test scores, so that differences in difficulty can be compensated for, but also considers initial aspects of the test design and construction (see, e.g., Holland and Rubin, 1982; von Davier et al., 2004; Dorans et al., 2007; von Davier, 2011; Kolen and Brennan, 2014).

In an observed-score equating (OSE) framework (von Davier, 2011), the equivalence of scores and equated scores are defined in terms of features of the test score distributions. For example, when equipercentile equating is used (see Eq. (1.2) and Section 1.4), the cumulative distributions of the random variables associated with the test scores from the two test forms are mapped onto each other such that the percentiles on one will match the percentiles on the other. If linear equating is used instead, the scores are defined to be interchangeable if they are at the same distance from the mean in standard deviation units. Hence, equipercentile and linear equating rely on a set of requirements, best practices for ensuring the quality of the data, and a set of methodologies for which the OSE framework was developed. In contrast, item response theory (IRT) models are mathematical models used to infer the ability of a test taker and to classify the test takers according to their ability. Linking the item parameters to adjust for form differences is inherent to the IRT model, and the equating requirements are covered by the IRT model assumptions (dimensionality, local independence, model fit, and parameter invariance). The requirement that the test forms should measure the same construct is also needed for IRT linking.

In the next section, we outline the main ideas behind equating using a formal mathematical statistics approach. This means that we will use the statistical definitions of random variables, parameters, sample spaces, and probability distributions when defining the equating transformations. As this book is primarily focused on the theoretical underpinnings of KE, we believe that a formal definition of the elements and steps involved in test score equating will help to provide a better understanding of the actual implementation of the methods. A similar approach has been provided in the book by González and Wiberg (2017) and in Chapter 1 in von Davier (2011).

1.2 Test Equating as a Statistical Modeling Problem

Test equating methods are statistical procedures that adjust for the test form differences in standardized testing programs, in which the test forms have been assembled according to the same specifications and have been administered under similar (standardized) conditions. The goal of test equating is to make it a matter of indifference for a test taker which test form he or she has taken. Nevertheless, the measurement process in standardized testing, which includes test equating, follows the same steps as a typical statistical modeling process (von Davier, 2011). The goal of (standardized) testing is to measure and compare the skills of the test takers regardless of which test form they took. In order to disentangle the differences in test forms and test takers' abilities, data from test takers are also collected according to specific designs and assumptions about the data-generating process are explicitly incorporated into the measurement model. From this perspective, test equating is only a part of the measurement process.

We believe that the statistical theory viewpoint of equating (von Davier, 2011; van der Linden, 2011; González and von Davier, 2013; Wiberg and González, 2016; González and Wiberg, 2017) is useful not only for understanding the different methods that will be described in the subsequent chapters, but also because it provides a basis for new research possibilities on equating.

1.2.1 Notation

In our discussion of test equating, we consider two test forms and denote them as X and Y. The test forms X and Y are composed by n_x and n_y items, and we assume that they are administrated to randomly sampled groups of test takers with sample sizes N_x and N_y, respectively. The scores obtained from test forms X and Y are denoted by X and Y, and are assumed to be random variables defined on the sample spaces $\mathcal{X} = \{x_1, x_2, \ldots, x_J\}$ and $\mathcal{Y} = \{y_1, y_2, \ldots, y_K\}$, and with distribution functions $F_X(x)$ and $F_Y(y)$, respectively. The probabilities of earning a certain score are defined as $r_j = \Pr(X = x_j)$ $(j = 1, \ldots, J)$ and $s_k = \Pr(Y = y_k)$ $(k = 1, \ldots, K)$, respectively. The actually observed score data will be assumed to be realizations of the random variables X and Y. We will also denote by N_j the number of test takers scoring x_j $(j = 1, \ldots, J)$ on test form X, and by N_k, the number of test takers scoring y_k $(k = 1, \ldots, K)$ on test form Y. If we are discussing only one test form, this will be assumed to be composed of n items and administered to N test takers.

When test forms are composed of binary-scored items (i.e, the test taker earns 1 if the item is correctly answered and 0 if it is wrong), a typical choice for a test score is to consider the total number of correct answers on the test, which is also known as the sum score. In this case, the sample spaces constitute the sets of all possible values for the scores in both test forms such that $\mathcal{X} = \{0, 1, 2, \ldots, n_x\}$ and $\mathcal{Y} = \{0, 1, 2, \ldots, n_y\}$. For instance, if test form X is composed of 30 binary-scored items and scores are assumed to be the total number of correct answers, then $\mathcal{X} = \{0, 1, 2, \ldots, 30\}$. Note, however, that other scoring rules, such as the ones used in IRT models, can produce different types of score scales (i.e., continuous rather than discrete). Here, the definitions will be given for what we call *observed scores*, which in this book will always be considered to be the total number of correct answers on a test.

Next, we introduce the general definition of a statistical model and discuss the basic concepts of parameters, parameter space, and sampling space in the context of test equating.

1.2.2 Equating as a Statistical Model

Statistical models assume that observed data are realizations of random variables following some probability distribution. Statistical models are used to learn about the data-generating mechanism. The random variables are assumed to follow a distribution F_θ that is indexed by a *parameter* θ defined on a *parameter space* Θ. A statistical model (e.g., Fisher, 1922; Cox and Hinkley, 1974; McCullagh, 2002) can be compactly written as

$$\mathcal{F} = \{\mathcal{X}, F_\theta : \theta \in \Theta\}, \tag{1.1}$$

where the family $\mathcal{F}$ is the collection of all probability distributions on $\mathcal{X}$ indexed by a parameter θ.

Under this approach, the sample spaces of two random variables representing the scores from two test forms will be seen as the score scales of such test forms, respectively. The equating problem will thus be reduced to finding an appropriate function to map the scores from one scale into their equivalents in the other scale. Mathematically, this means that a function has to be defined that takes values in $\mathcal{X}$ and gives as a result a value on $\mathcal{Y}$. The following definition is general for any equating transformation.

Definition 1. *Let $\mathcal{X}$ and $\mathcal{Y}$ be two score scales. A function $\varphi : \mathcal{X} \mapsto \mathcal{Y}$ that satisfies the symmetry requirement will be called an equating transformation.*

The symmetry requirement ensures a symmetric equating, in the sense that if $\varphi : \mathcal{X} \mapsto \mathcal{Y}$ equates X to Y, then $\varphi^* : \mathcal{Y} \mapsto \mathcal{X}$ equates Y to X with $\varphi^* = \varphi^{-1}$.

The definition of observed-score test equating given by Braun and Holland (1982) is more specific, introducing the equipercentile as the functional form for φ (Angoff, 1971). This definition has been used by several authors (e.g., von Davier et al., 2004; Dorans et al., 2007; von Davier, 2011; González and Wiberg, 2017) and will be further used in this book.

Let x and y be the quantiles in the distributions of tests scores X and Y for an arbitrary common cumulative proportion p of the population, i.e., $x(p) = F_X^{-1}(p)$ and $y(p) = F_Y^{-1}(p)$. It follows then that for every $p \in [0, 1]$, with the requirement that $F_Y(y) = F_{\varphi(x)}(y)$, an equivalent score y on test Y for a score x on X can be obtained as

$$y = \varphi(x) = F_Y^{-1}(F_X(x)). \tag{1.2}$$

Although Eq. (1.2) can be used in general for the mapping of any two samples or distributions of random variables (see, e.g., Wilk and Gnanadesikan, 1968), the function φ is known in the equating literature as the equipercentile transformation (Angoff, 1971).

From Eq. (1.2), it follows that an estimate of the equating transformation φ can be obtained by estimating the score probability distributions F_X and F_Y. A natural estimate for each of these functional parameters is the empirical cumulative distribution function (ECDF). However, ECDFs are discrete step functions rendering the corresponding inverses unavailable in most cases. Moreover, when $\mathcal{X} \subset \mathbb{N}$ and $\mathcal{Y} \subset \mathbb{N}$, which is most typically the case in operational testing in which test scores correspond to the total number of correct answers on the test, the score cumulative distribution functions (CDFs) F_X and F_Y will be discrete because the possible values that sum scores can take are consecutive integers[1]. The problem of discreteness thus remains. Recently, González et al. (2024) showed that when using generalized inverses (Embrechts and Hofert, 2013), discrete score distributions actually do not present a problem to obtaining the inverse CDFs and computing an equating transformation, although the actual obtained φ function might not fulfill the symmetry requirement. In common practice, estimates of φ are based on continuous approximations of the originally discrete distributions F_X and F_Y. In the equating literature, this practice is called *continuization*, and it requires one to actually "continuize" the discrete score distributions in order to properly use Eq. (1.2) for equating (van der Linden, 2019). Typical continuization methods used in equating include linear interpolations, kernel smoothing techniques, and methods that use the polynomial log-linear models (see Wang, 2011).

For obvious reasons, the methods for the continuization of test score distributions play a central role in this book. We will focus on the kernel smoothing techniques for continuizations, which in Chapter 5 will be seen as the result of convolution operations between the discrete score probability distributions and continuous distribution functions.

In practice, test score data are needed to estimate φ. The equating transformation is thus a functional parameter in the equating model that needs to be evaluated in terms of accuracy, bias, and possibly other features. How to evaluate equating transformations is described in Chapter 8.

An important aspect in the definition of φ is the fact that it has to be valid for a certain population. As such, it needs to be defined on a common population of interest, which is not always trivial for some data collection strategies. In the next section, we briefly describe the target population where

---

[1]In this book we use the convention $\mathbb{N} = \{0, 1, 2, \dots\}$

the equating transformation is defined and describe different data collection designs.

1.3 The Target Population and Data Collection Designs

The selection of items on a test form and the selection of the samples of test takers for the analyses are designed to provide relevant information about the test takers' abilities. In this section, the target population, equating samples, and several common types of data collection designs for equating are briefly discussed.

1.3.1 The Target Population

Any assessment is developed to provide the maximum information about test takers from a target population $\mathcal{T}$. We want the assessment to be invariant across subgroups in $\mathcal{T}$, for example with respect to gender or ethnicity. Similarly, an equating of test forms of this assessment is also subpopulation invariant across subgroups in $\mathcal{T}$ (see requirements in Section 1.2). Usually, this requirement tends to hold well for most traditional standardized assessments, for which the target population is clearly defined and does not change significantly over time (see von Davier and Wilson, 2007). When subgroups of test takers diverge over time in their levels of the skill measured by the test, the equating function might, however, become group-dependent. As will be described in Section 1.3.6, common items (i.e., an anchor test) can be useful when equating test scores under this scenario. In large-scale assessments usually only one or a few subgroups are given an anchor test due to test security. Recent research has concluded that in which subgroup the anchor test is given in comparison to the target population may impact the equating transformation (Laukaityte and Wiberg, 2022, 2024).

In the past decade, the impact of demographics on the equating results over time has been studied from various perspectives (Duong and von Davier, 2013; Lee and von Davier, 2013; Li et al., 2012; Qian et al., 2013). New methods of equating that incorporate test takers' covariates have also been proposed (Wiberg and Bränberg, 2015; Wallin and Wiberg, 2017, 2019; Sansivieri and Wiberg, 2017, 2019). It has also become evident that it is important to monitor the groups of test takers in connection to ancillary information about them (Wiberg and von Davier, 2017). Further, it is important to monitor the equating transformation over time for undesirable trends and to consider including adjustments in order to maintain the validity of the testing program (Wiberg, 2017). Finally, the issue of measurement invariance across subgroups and subpopulation invariance of the equating transformations are important topics to consider.

A different view on what constitutes the target population, especially when an anchor test is used, is given in San Martín and González (2022). These authors formally specify the statistical model underlying the NEAT design (see Section 1.3.6) using a fully probabilistic approach. In this model, the sample space is the actual population of interest and corresponds to the union of the index sets labeling the two mutually exclusive groups of test takers who were exposed to the test forms X and Y.

1.3.2 Data Collection Designs

The observed score data can be thought of as a realization of a random variable that represents the score of a test taker belonging to a certain group in a certain population. Although in principle it is possible to obtain data from any number of different groups of test takers within any number of different populations, the exposition that follows will consider only one target population (say, college applicants), two populations (say, cohorts of college applicants), and two groups within each population (say, the students who took the test forms).

Figure 1.1 shows a schematic representation of the different data collection designs that are considered in this book. In this figure, the rectangle in the middle represents the target population, $\mathcal{T}$, and the circles with solid lines represent two different populations, $\mathcal{P}$ and $\mathcal{Q}$. Circles with pointed lines are randomly sampled groups either from the target population, or from the two different populations. The squares represent test forms (X, Y, and A) and covariates (C). Finally, the arrows are used to indicate both that groups of test takers are sampled from one or two different populations, and that test forms have been administered to one or two groups of test takers.

In the following subsections, we use Figure 1.1 to describe the data collection designs that usually comprise building blocks for the operational equating. The way the data are collected is relevant for the theory of test equating because the design features need to be incorporated into the generalized equating framework. These designs have been extensively described previously in the literature, so the interested reader is referred to von Davier et al. (2004), Dorans et al. (2007), von Davier (2011), Kolen and Brennan (2014), and González and Wiberg (2017) for more details.

1.3.3 Equivalent Groups Design

In the equivalent groups (EG) design, two independent groups of test takers (G_1 and G_2) are sampled from a common target population $\mathcal{T}$, and each group is administered one of the two test forms. The obtained data are two independent samples of scores x_i ($i = 1, \ldots, N_x$) and $y_{i'}$ ($i' = 1, \ldots, N_y$).

1.3.4 Single Group Design

In the single group (SG) design, one group (G) of test takers from a common target population $\mathcal{T}$ is administered both test form X and test form Y, and

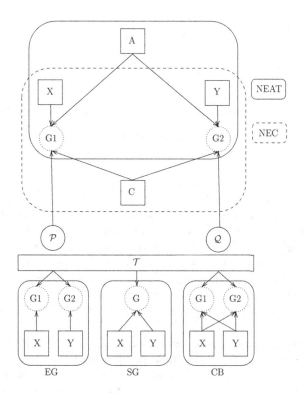

FIGURE 1.1: Different equating designs (EG = Equivalent Groups, SG = Single Groups, CB = Counterbalanced, NEAT = Nonequivalent groups with Anchor Test, and NEC = Nonequivalent groups with Covariates) when equating from test form X to test form Y. A is the anchor test form and C represents covariates. G, $G1$, and $G2$ are groups from population $\mathcal{P}$ and/or $\mathcal{Q}$, and $\mathcal{T}$ is the target population.

the resulting data are thus a sample of a bivariate score vector (x_i, y_i), for test takers $i = 1, \ldots, N$.

1.3.5 Counterbalanced Design

In the counterbalanced (CB) design, two independent groups of test takers are sampled from the same target population $\mathcal{T}$. Both groups are administered both test forms in different orders. We use the subindexes 1 and 2 to denote the order in which the tests were administered. Assume that N_{G_1} and N_{G_2} test takers are sampled in groups G_1 and G_2, respectively. The obtained data are two independent bivariate vectors (x_{1i}, y_{2i}) $(i = 1, \ldots, N_{G_1})$ and $(x_{2i'}, y_{1i'})$ $(i' = 1, \ldots, N_{G_2})$, where in this case Group 1 was first administered test X and then Y, whereas Group 2 was first administered test Y and then X. It is easy to see from Figure 1.1 that two EG designs and two SG designs are implicitly embedded in a CB design. A discussion of different possibilities for using the data in a CB design is given in von Davier et al. (2004).

1.3.6 Non-Equivalent Groups with Anchor Test Design

In the non-equivalent groups with anchor test (NEAT) design, two groups are independently sampled, each from one of two populations, $\mathcal{P}$ and $\mathcal{Q}$. Each sampled group is administered either test form X or test form Y, and both groups are administered a common anchor test form A composed of n_a items. Let A be the random variable defined on $\mathcal{A} = \{a_1, a_2, \ldots, a_L\}$ representing the anchor test score and $a_l = \Pr(A = a_l)$ $(l = 1, \ldots, L)$. The obtained data are score vectors for N_x and N_y test takers answering both the anchor test form A and either of the test forms X or Y, respectively. More precisely, the data consists of the two samples of score vectors (x_i, a_i) $(i = 1, \ldots, N_x)$ and $(y_{i'}, a_{i'})$ $(i' = 1, \ldots, N_y)$. Anchor scores can be either *internal*, i.e., the anchor score is included in the total reported sum score, or external, i.e., the anchor score is only used for equating purposes and is excluded in the reported sum score.

An anchor test should have a similar content as the regular test. Research has shown that it is preferable, from an equating perspective, if the anchor test is a miniature version of the complete test, i.e., a *minitest* (e.g., Kolen and Brennan, 2014). Note, however, that there are some criticisms of this recommendation (e.g., Sinharay et al., 2012), and there have been suggestions to instead use a *miditest*, i.e., a test of average difficulty but of less spread. Recently, Laukaityte and Wiberg (2024) examined both how the anchor test should be designed in terms of discrimination and difficulty compared with the regular test, and to which groups it is preferable to administer the anchor test if there is a choice. In line with previous research, they concluded that using an anchor test that is easier than the regular test, as well as anchor items that have more spread in difficulty compared with the regular test, had a negative effect on the standard error of equating (SEE); see Section 8.1.4 for a definition of SEE. A more important result was that for equating purposes, it is better to give an anchor test to an average ability group compared with a high-performing ability group as the former resulted in smaller SEE. In some situations, however, we may not have access to an anchor test, although there might be additional collected information about the test takers. In these circumstances, one can instead use the non-equivalent groups with covariates design, as described below.

1.3.7 Non-Equivalent Groups with Covariates Design

In the non-equivalent groups with covariates (NEC) design, information about the test takers in the form of covariates, C, is used in order to obtain a more precise equating (Wiberg and Bränberg, 2015). The underlying assumption here is that the groups' differences can be described accurately by the covariates. The NEC design is similar to the NEAT design, although the covariates are used instead of an anchor test. In the NEC design, two groups are independently sampled from the populations $\mathcal{P}$ and $\mathcal{Q}$, and each group is administered either test form X or test form Y. New equating methods have been proposed for the NEC design by, for example, Sansivieri and Wiberg (2017, 2019). Further, Albano and Wiberg (2019) have evaluated the accuracy when covariates are used in the equating with respect to anchor type, sample size, test length, and ability differences. They concluded that the equating transformation in most conditions was more accurate when including covariates.

Wallin and Wiberg (2017, 2019) proposed using propensity scores to incorporate covariates in the equating. Propensity scores specify the conditional probability of being assigned to a specific treatment (in this case, a specific test form) given a covariate vector. Recently, Wallin and Wiberg (2023) examined model misspecification and robustness of the equating transformation when using propensity scores in the NEC design. Their overall conclusions were that failing to include an important covariate in the propensity score estimation model introduced bias, but failing to include a second-order term in the propensity score model did not produce significant bias. In practice, this means that when using a NEC design, it is extremely important to incorporate all important covariates in the model. Other ways to incorporate covariate information in equating include using the background information to form pseudo-equivalent groups before conducting the equating (Haberman, 2015).

1.4 Equating Transformations

Definition 1 does not restrict the shape of φ as long as it produces a symmetric equating. The equipercentile equating transformation in Eq. (1.2), for instance, is a curvilinear function that is easily shown to produce a symmetric equating, as shown in Section 7.3.

Another popular equating transformation is a linear function, which can be shown to be a particular case of Eq. (1.2) in the case when the score distributions are location-scale families (see Section 7.4). If X and Y have means μ_X and μ_Y, and standard deviations σ_X and σ_Y, then the linear equating transformation under the EG design is defined as

$$\varphi(x; \boldsymbol{\theta}) = \mu_Y + \frac{\sigma_Y}{\sigma_X}(x - \mu_X). \tag{1.3}$$

In this case, we have a completely parametric estimator, as the parameter $\boldsymbol{\theta} = (\mu_X, \mu_Y, \sigma_X, \sigma_Y) \in \Theta = \mathbb{R} \times \mathbb{R} \times \mathbb{R}^+ \times \mathbb{R}^+$ fully characterizes the score distributions that generate the data. Sample means and variances can be directly estimated from the observed data. Note that the parameter vector $\boldsymbol{\theta}$ might contain other parameters when the linear equating function is derived under other data collection designs. For instance, we have seen that bivariate vectors of score data are obtained for the NEAT design, and thus covariance parameters will also play a role in estimating φ. For a formal definition of the linear transformation when using covariances in the NEAT design, refer to von Davier and Kong (2005) and González and San Martín (2024).

In general, estimators of φ can be parametric (linear case), semi-parametric, or nonparametric estimators. In the nonlinear cases, the continuization method defines the statistical inference approach used (González and von Davier, 2013; González and Wiberg, 2017). For instance, the equipercentile transformation is estimated using observed score data $x_1, \ldots, x_{N_x} \sim F_X$ and $y_1, \ldots, y_{N_y} \sim F_Y$, where F_X and F_Y are the CDFs of X and Y, respectively. Since no particular parametric family of score distributions is specified, the data-generating mechanism can be completely described by the two CDFs resulting in a nonparametric estimator. This estimator is impacted by the discreteness of the distributions, and traditionally, linear interpolation has been used to continuize the obtained discrete distributions.

Semiparametric equating estimators of φ include the kernel equating function described in the next subsection as well as equating transformations used in local equating (van der Linden, 2011), and combinations of kernel and local equating (Wiberg et al., 2014). These equating transformations are described thoroughly in Chapter 7.

1.4.1   The Kernel Equating Function

Kernel equating (KE) (von Davier et al., 2004) can be viewed as a semi-parametric equating estimator, as both finite-dimensional parameters and distribution functions are involved in the estimation of φ (González and Wiberg, 2017).

The parametric part consists of the score probability parameters $r_j = \Pr(X = x_j)$, and $s_k = \Pr(Y = y_k)$ with x_j and y_k taking values in $\mathcal{X}$ and $\mathcal{Y}$, respectively. Typically, these are viewed as parameters of a multinomial distribution. The nonparametric part comes in the estimation of the score distribution functions using kernel smoothing techniques. The equating transformation in KE is defined as an equipercentile equating function, in which the test scores' CDFs have been obtained by continuizing the test score distributions using a kernel method. The mathematical form of φ is

$$\varphi(x; \boldsymbol{\theta}) = F_{h_Y}^{-1}(F_{h_X}(x; \boldsymbol{r}), \boldsymbol{s}),$$

with $\boldsymbol{\theta} = (\boldsymbol{r}, \boldsymbol{s})$ and where $\boldsymbol{r}$ and $\boldsymbol{s}$ are J-dimensional and K-dimensional vectors with elements r_j and s_k, and h_X and h_Y are bandwidth parameters controlling the smoothness in the continuization.

1.5 Evaluating the Equating Transformation

When conducting equating, there is always a risk that we get systematic errors, random errors, or both. It is therefore important to evaluate the equating transformation with different measures and methods. Different methods and measures have been proposed over the years to evaluate the equating transformation, and in this book, we distinguish between equating-specific measures and statistical measures. The former are measures developed explicitly for evaluating the equating transformation and are most likely not suitable for evaluating other statistical estimators. Statistical measures are measures used in general statistical inference for evaluating estimators. In this book, we will describe some statistical measures that are suitable to use when evaluating an equating transformation. Both these groups of measures and the methods therein will be described in detail in Chapter 8, and examples with real data are given in Chapter 9 for the EG design and in Chapter 10 for the NEAT design.

In the next two sections, we briefly describe the practical elements of this book: the software packages and the data used for the illustrations.

1.6 R Packages Used in This Book

Different equating methods are described throughout this book, and illustrations are given in Chapters 9 and 10. A unique feature of this book compared with the von Davier et al. (2004) book is that **R** code is provided for the described equating methods. Essentially, we have used the **R** packages **kequate** (Andersson et al., 2013), which implements both KE and IRT KE, and **SNSequate** (González, 2014), which implements both traditional and KE methods.

Other **R** packages are used for models and methods that are helpful in particular stages for some of the equating methods. When implementing presmoothing with discrete kernels, we utilized the **R** package **SNSequate**. The functions within this package are built on top of functions from the **ake** package (Wansouwé et al., 2022). To perform IRT KE, the packages **ltm** (Rizopoulos, 2006) and **mirt** (Chalmers, 2012) are used to estimate the IRT models of interest. Finally, **psych** (Revelle, 2023) is used to examine the IRT model fit.

1.7 Empirical Examples

In the last part of the book (Chapters 9 and 10), we illustrate some of the described methods in the EG and NEAT designs with real data examples. The first example comes from a private national evaluation system in Chile called SEPA (Sistema de Evaluación del Progreso en el Aprendizaje; System of Assessment Progress in Achievement) administered by the measurement center MIDE UC at Pontificia Universidad Católica de Chile. SEPA consists of tests specifically designed to assess achievement in students from first to eleventh grade in the subjects of Language and Mathematics.

The second example, which we called ADM, contains two administrations of a college admissions test, which aim to select the test takers who are most likely to succeed in university. The test consists of a verbal section and a quantitative section, which are equated separately with a NEAT design. The ADM data set thus also contains an anchor test. In addition, simulated data sets are used in the book.

1.8 Summary and Overview of the Book

The book is divided into three parts. The first part, which includes this chapter and Chapter 2, introduces the definitions and the basic elements of the generalized kernel equating (GKE) framework. The second part, which consists of Chapters 3–8, is focused on the GKE framework. More specifically, it involves presmoothing, estimation of score probabilities, the continuization of the test score distributions through the use of convolutions, bandwidth selection for the kernels, the equating transformation, and the evaluation of the equating transformation. The third part of the book (Chapters 9 and 10) describes how some of the methodology presented in the previous chapters can be applied to empirical examples, what practical considerations need to be made, and how **R** code can be used. More specifically, the book contains the following.

In Chapter 2, *Kernel Equating*, we review the KE framework (von Davier et al., 2004; von Davier, 2011, 2013). Current and new methods used in each of the five steps – presmoothing, estimation of score probabilities, continuization, computing the equating transformation, and evaluation of the equating – are briefly introduced in order to define the GKE framework. A detailed description of the methods in each of these steps is postponed to the following chapters.

In Chapter 3, *Presmoothing*, different presmoothing strategies, including parametric statistical models, nonparametric methods, and mixture distributions, are explored together with how they are evaluated. We review the traditional log-linear models for presmoothing, but also describe discrete kernels and methods based on mixture distributions.

In Chapter 4, *Estimating Score Probabilities*, we describe the estimation of the score probabilities depending on which data collection design and presmoothing modeling method have been used. In this chapter, we review the material on design functions (von Davier et al., 2004) and introduce new methods for the estimation of score probabilities, including how IRT models are used in connection with KE.

In Chapter 5, *Continuization*, the Gaussian kernel continuization of the score distributions, which has been the standard selection in KE, is further extended by using the theory of convolutions. Gaussian continuization is shown to be a particular case in this framework. The theoretical properties of the resulting continuous distributions are described not only for the Gaussian, but also for the uniform, logistic, Epanechnikov, and adaptive kernels.

Chapter 6, *Bandwidth Selection*, emphasizes the importance of the bandwidth selection in KE. The chapter extends the traditional penalty function method to select the bandwidth parameter in KE (von Davier et al., 2004), introducing a number of new bandwidth selection methods that have emerged during the past few years. Some of these methods are the rule-based method (Andersson and von Davier, 2014), a likelihood-based cross-validation (Liang and von Davier, 2014), double smoothing (Häggström and Wiberg, 2014), and the penalty cross-validation approach (Wallin et al., 2021). A new bandwidth selection method, which is based on the test score CDFs instead of the probability density functions (PDFs), is also suggested.

In Chapter 7, *Equating*, the KE equating transformation and variants of it (e.g., transformations used for chained and local equating) are described under the GKE framework. We also discuss how linear versions of the equating transformation can be obtained under the GKE framework.

Chapter 8, *Evaluating the Equating Transformation*, comprises a review and illustration of the approaches for the evaluation of the equating transformation. We also provide recommendations on how to properly design a simulation study in KE for the purpose of conducting comparisons among different equating transformations.

In Chapter 9, *Examples under the EG Design*, real data sets are used to exemplify some of the methodologies given in this book for the EG design. We guide the reader through each of the GKE steps, provide code and recommendations, and discuss our approaches in the context of the empirical example.

In Chapter 10, *Examples under the NEAT Design*, similarly as for the EG design, real data sets are used to exemplify the methodologies given in this book under a NEAT design. Again, we guide the reader through each of the GKE steps, provide code and recommendations, and discuss our approaches in the context of the empirical example.

The book concludes with two appendices. Appendix A contains instructions on the installation of **R**, a list of **R** packages with version number used within the book, as well as a brief description of how to read in test data. Appendix B describes which **R** packages can be used to perform the different steps in GKE and also refers to where different examples can be found.

2

Kernel Equating

In this chapter, we review the KE framework (von Davier et al., 2004), which has traditionally been conceptualized into five steps.

The first step is presmoothing, in which we account for potential irregularities in the score probability distributions. Moving on to the second step, we map the score probabilities that define score distributions into marginal score probabilities for each of the equating designs. These mapped score probabilities serve as weights in the third step, in which we use kernel techniques to transform discrete score distributions into continuous ones. In the fourth step, we compute the equating transformation, and in the final, fifth step, we calculate the standard error of equating.

The primary focus of this chapter is to provide a brief overview of the five steps within the KE framework, and explain their general purposes. We will not only discuss the methods originally proposed for each step but also introduce several new alternatives that are part of the generalized kernel equating (GKE) framework. For a more detailed description of the methods employed in each of the five steps, please refer to the consecutive chapters in this book.

2.1 Introduction

When KE was first proposed in Holland and Thayer (1989), only three steps were considered: estimation, continuization, and equating. Estimation was used to refer to the process of fitting a statistical model to the discrete score distributions. Later, this step was renamed presmoothing, and two more steps were added: estimation of score probabilities, and calculation of the standard error of equating (Holland et al., 1989; von Davier et al., 2004). In what follows, we describe and give a brief account of all these steps, postponing the details of the methods used in each of them for later chapters. We do not focus on particular models or methods that are used in each of these steps. Rather, the exposition is intended to be as general as possible, reflecting the fact that KE can be framed in a *generalized kernel equating framework*.

Suppose that we are interested in equating two test forms, X and Y, and that the score data have been collected under the EG design. Let $x_j \in \mathcal{X}$ and $y_k \in \mathcal{Y}$ denote the possible score values that the random variables X and

Y can take with $j = 1, \ldots, J$ and $k = 1, \ldots, K$, respectively. The associated cumulative distribution functions (CDFs) of the score probability distributions are defined as

$$F_X(x) = \Pr(X \leq x) = \sum_{j, x_j \leq x} r_j, \qquad (2.1)$$

$$F_Y(y) = \Pr(Y \leq y) = \sum_{k, y_k \leq y} s_k, \qquad (2.2)$$

where $r_j = \Pr(X = x_j)$ and $s_k = \Pr(Y = y_k)$ are the score probabilities that define the distributions, and $x, y \in \mathbb{R}$. Because both $F_X(x)$ and $F_Y(y)$ are discrete CDFs, an equipercentile equating transformation, as defined in (1.2), cannot be computed straightforwardly. The continuization step turns these discrete CDFs into the continuous $F_{h_X}(x)$ and $F_{h_Y}(y)$, where h_X and h_Y are parameters used to control the degree of smoothness in the continuization. The corresponding density functions $f_{h_X}(x)$ and $f_{h_Y}(y)$ are found by differentiating the continuous CDFs of X and Y with respect to x and y, respectively. Moreover, in Chapter 5 we will see that these continuous distributions also depend on the score probability parameters r_j and s_k and they can be rewritten as $F_{h_X}(x; \boldsymbol{r})$ and $F_{h_Y}(x; \boldsymbol{s})$, with $\boldsymbol{r} = (r_1, \ldots, r_J)^t$ and $\boldsymbol{s} = (s_1, \ldots, s_K)^t$. From these definitions, the equating transformation can be obtained as described in Section 1.4.1, using (1.2) as follows:

$$\varphi(x) = F_{h_Y}^{-1}(F_{h_X}(x, \boldsymbol{r}), \boldsymbol{s}). \qquad (2.3)$$

Next, the collected score data are used to obtain the following estimate: $\hat{\varphi} = F_{h_Y}^{-1}(F_{h_X}(x, \hat{\boldsymbol{r}}), \hat{\boldsymbol{s}})$. Although the sample relative frequencies can be used as estimates $\hat{r}_j$ and $\hat{s}_k$, different methods are used in the presmoothing step to fit the score data and find better estimates of the score probabilities. The functional parameter estimate $\hat{\varphi}$ is evaluated in the last step.

Figure 2.1 shows a schematic representation of the five steps in KE for this example. Note that as score data have been collected under an EG design, the KE process starts with two independent univariate score distributions that are separately smoothed (Step 1). However, if score data come from bivariate distributions, as is the case, for example, in the SG and NEAT designs, bivariate score distributions are to be presmoothed. Also note that the equating step (Step 4) that follows after continuization (Step 3) always considers the composition of two *univariate* score distributions[1] that depend on marginal score probabilities. Performing Step 2 in the case of bivariate distributions will lead to two marginal vectors of score probabilities which are mapped from (smoothed) bivariate distributions. For the EG design, such mapping is just the identity function, as univariate instead of bivariate distributions are considered.

---

[1] An exception occurs when the equating transformation is obtained as a chain of composition of more than two univariate CDFs (see Chapter 7).

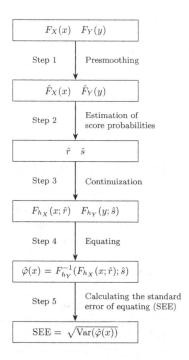

FIGURE 2.1: The five steps in KE under the EG design.

2.1.1 The Five Steps in KE

In the original formulation of KE (von Davier et al., 2004), only a few methods were used in each of the five steps, as summarized below:

- *Step 1. Presmoothing*: Presmoothing the observed score distributions with polynomial log-linear models.

- *Step 2. Estimating score probabilities*: Use of the design function (DF) to estimate score probabilities from the distributions obtained in Step 1.

- *Step 3. Continuization*: Use of the Gaussian kernel and a bandwidth selection method, which minimizes a penalty function in order to make the discrete score distributions continuous.

- *Step 4. The equating transformation*: Calculating the equating transformation by using continuized marginal score distributions.

- *Step 5. Calculating the standard error of equating (SEE)*: Evaluating the equating transformation by calculating the standard error of equating (SEE), standard error of equating differences (SEED), and percent relative error (PRE).

The content of these five steps is included in more detail in different parts of this book. What is important to note is that all the elements involved in ˴ the KE transformation are related to some of the five steps listed above. In the following section, these steps are described in general terms together with new methods that can be used within each of them.

2.2   The Generalized Kernel Equating Framework

We propose the GKE framework, which is a generalization of the KE framework. By proposing a framework, we give a clear structure that allows us to include existing methods but leaves room for including new methods in the future. We view the KE approach as a set of steps in which there are a large number of different alternatives. Similar to KE, the GKE framework consists of a five-step procedure that one proceeds through when conducting test score equating. The largest difference from KE is that the GKE framework allows for more flexibility and more choices than in KE. The GKE framework stipulates that we will have several methods that we can choose between in each of the five steps that are used to conduct an equating. Once we have chosen a suitable method within a step, there are a number of different choices that can be made within a method depending on, for example, the data collection design used.

The GKE framework can also be viewed as a large tree, in which the different branches represent the different steps. On each branch can be a large number of methods or leaves. This also means that the framework is flexible so that new methods can easily be included in the different steps (or branches).

It is important to emphasize that although we describe a large number of methods within the GKE framework in this book, there are more methods that can be included in each step. Thus, the GKE framework is flexible and not limited to the current existing methods described in this book. We want to emphasize that we expect the number of useful methods to increase within every step over the coming years. A summary of methods that can be used within each step and which are later described in this book is provided in Table 2.1, and a detailed description of each of them is given in the following chapters of this book.

In the following subsections, we will, for each of the five steps of GKE, describe the step in general terms and leave the details to the following chapters. We will briefly describe different methods that have been used in the past with KE as well as methods we propose to use in the future. All these methods will be described in detail in Chapters 3–8.

TABLE 2.1: A few methods considered in the GKE framework.

Step	Methods
1. Presmoothing	Polynomial log-linear models
	Discrete kernels
	Beta4 models
	IRT models
2. Estimation of score probabilities	Design functions
	Lord-Wingersky algorithm
	Poisson's binomial distribution
	Normal approximations
3. Continuization	Gaussian kernel
	Logistic kernel
	Uniform kernel
	Epanechnikov kernel
	Adaptive kernels
	Percentile rank method
3. Bandwidth Selection	Minimizing a penalty function
	Cross-validation methods
	Double smoothing
	Rule-based method
	CDF-based method
4. Equating	Equipercentile equating
	Linear equating
	Chained equating
	IRT equating
	Local equating
5. Evaluation	Difference that matters (DTM)
	Percent relative error (PRE)
	Standard error of equating (SEE)
	Equating indices
	Standard error of equating difference (SEED)
	Bias
	Mean squared error (MSE)
	Root mean squared error (RMSE)
	Standard error (SE)
	Cumulants

2.2.1 Presmoothing

The goal of presmoothing is to reduce sample irregularities seen in the empirical score distributions that arise due to sampling error. This can be achieved by using statistical models and methods to fit the data. In KE, polynomial log-linear models were typically used for the presmoothing step. In the GKE framework, we propose using other methods for presmoothing as well. In particular, we distinguish between three classes of methods: i) parametric statistical models, ii) nonparametric methods, and iii) methods based on mixture models.

In the first class, presmoothing is achieved by fitting a parametric statistical model that relates the score probabilities to the model parameters. Formally, if p represents the score probabilities under one of the data collection designs in equating, and $\mathcal{M}(\boldsymbol{\theta})$ is a model parameterized by $\boldsymbol{\theta}$, then

$$p = \mathcal{M}(\boldsymbol{\theta}) \tag{2.4}$$

is fitted to the data, and the model parameter estimates, $\hat{\boldsymbol{\theta}}$, are used to determine an estimation of p. Note that under the EG design, the components of the vector p are $r_j = \Pr(X = x_j)$, $j = 1, \ldots, J$, whereas under the SG and NEAT designs they are $p_{jk} = \Pr(X = x_j, Y = y_k)$ and $p_{jl} = \Pr(X = x_j, A = a_l)$, respectively. In the case of the CB design, the two bivariate probabilities are specified as $p_{(12)jk} = \Pr(X_1 = x_j, Y_2 = y_k)$ and $p_{(21)jk} = \Pr(X_2 = x_j, Y_1 = y_k)$, respectively. Similar score probabilities can be defined for Y.

The most used statistical models for presmoothing in KE are the family of polynomial log-linear models (Holland and Thayer, 2000). To perform presmoothing with log-linear modeling, a polynomial log-linear model is fitted to the observed proportions in the raw data. This was the presmoothing method originally proposed in KE. Details of this method are given in Chapter 3 and can also be found in Holland and Thayer (2000), von Davier et al. (2004), and González and Wiberg (2017).

In the second class, which contains nonparametric methods, we suggest in Chapter 3 a nonparametric presmoothing alternative, which makes use of discrete kernel estimators (Rajagopalan and Lall, 1995) instead of parametric statistical models.

In the third class, where we consider methods based on mixture models, we have a model introduced by Lord (1965) that relates observed and true scores. Among other features, this mixture model is capable of estimating the score frequency distribution and is thus suitable for presmoothing. This model has been successfully applied for presmoothing in equipercentile equating by Kolen and Brennan (2014), who called this approach the beta4 presmoothing procedure. In Chapter 3, we will use beta4 models for presmoothing, which, to the best of our knowledge, have not been used yet in KE. Other statistical models belonging to the mixture models class and that have been used for presmoothing are those from the family of IRT models. Under this approach, score probabilities are obtained based on either binary or polytomous IRT models depending on the nature of the data. In Chapter 7, the use of IRT models is discussed for the estimation of score probabilities p (Andersson and Wiberg, 2014, 2017; Andersson, 2017).

Note that polynomial log-linear models, beta4 models, and IRT models all make use of parametric models to fit the score frequency distributions, while in the second class estimates of the score distributions are obtained using kernel techniques.

An important feature, however, for all presmoothing methods is that they should be flexible enough to smooth either univariate or bivariate estimated probabilities, depending on the design that has been used for the collection of data. Furthermore, whatever the presmoothing method adopted, a good fit of the model to the score data is mandatory. In Chapter 3, different measures of goodness of fit will be introduced to select the best model to use for presmoothing. Presmoothing is not mandatory if the samples are sufficiently large, and if presmoothing has not been used, we can instead follow the ideas of Moses and Holland (2007) and use the sample estimates of r and s.

Note that even though we propose several different presmoothing methods, it is possible that other presmoothing methods can be useful to include within the GKE framework in the future. This flexibility of the GKE framework is similar for the other four steps.

2.2.2 Estimating Score Probabilities

We have seen in Chapter 1 that it is possible to model score data separately (i.e., two univariate distributions in the EG design) or jointly (i.e., one bivariate distribution in the SG, or two independent bivariate distributions in the CB, NEAT, or NEC designs). The computation of the equating transformation (2.3), however, makes use of marginal univariate score probabilities that act as weights in the kernel estimation of the score distributions (see Figure 2.1 and the discussion in Section 2.2.1). To obtain the marginal score probabilities, the design function $\mathbf{DF}(\cdot)$ (DF, see Section 4.1) is used to map the population score probabilities p into the marginal score probabilities $r = (r_1, r_2, \ldots, r_J)^t$ and $s = (s_1, s_2, \ldots, s_K)^t$. The equating transformation thus can be written in the form

$$\varphi(x) = \varphi[x; \mathbf{DF}(p)] = \varphi(x; r, s). \tag{2.5}$$

In other words, the equating transformation depends on the score values and the design function that maps the population score distributions into marginal score probabilities. Depending on the data collection design, different DFs will be used. Explicit definitions of the DFs for the EG, SG, CB, and NEAT designs are given in Chapter 4, but can also be found in von Davier et al. (2004), and González and Wiberg (2017). In the GKE framework, all these DFs are included together with DFs for the NEC design. Explicit definitions of the DFs for different NEC designs are found in Chapter 4, although some of them can also be found in Wiberg and Bränberg (2015), González and Wiberg (2017), and Wallin and Wiberg (2019).

Once a model has been fitted in the presmoothing step, the estimated score probabilities $\hat{p}$ are passed to the DF to obtain estimated marginal score probabilities $\hat{r}$ and $\hat{s}$. To avoid confusion, in what follows we will refer to $\hat{p}$ as the *estimated probabilities*, whereas the $\hat{r}$ and $\hat{s}$ will be called *estimated score probabilities.* Note that in KE, the estimated probabilities have been typically obtained from log-linear models. In the GKE framework, these can also be obtained from other statistical models and methods, such as discrete kernel estimators, beta4 models, and IRT models. For example, to obtain estimated probabilities from an IRT model, an iterative algorithm due to Lord and Wingersky (1984) is often used. How to obtain estimated score probabilities when different presmoothing models have been used is discussed in Chapter 4.

2.2.3   Continuization

In order to properly estimate the equating transformation defined in Eq. (2.3), a common practice in equating is to use continuous approximations of the original discrete score distributions F_X and F_Y. This practice is referred to as *continuization* in equating terminology. The continuized score distributions are then used to obtain φ.

Different alternatives exist for obtaining continuous approximations of score distributions and the following methods have been used in equating: linear interpolation, continuized polynomial log-linear modeling (Wang, 2011), kernel smoothing techniques (Silverman, 1986; von Davier et al., 2004), and exponential families (Haberman, 2011). However, not all of them are based on the use of a kernel function to approximate the score distributions. Because the focus in this book is on KE, only the kernel continuization is considered in detail in Chapter 5. Kernel continuization is shown to result from a convolution operation between the discrete score probability distribution function and a continuous probability distribution function. The originally proposed continuous distribution in KE was the Gaussian, leading to the Gaussian kernel continuization (von Davier et al., 2004). In Chapter 5, we will describe this option together with other options for continuization included in the GKE framework, such as logistic, uniform (Lee and von Davier, 2011), Epanechnikov, and adaptive kernels (González and von Davier, 2017).

2.2.3.1   Bandwidth Selection

The kernel estimators used in the GKE framework, which were briefly mentioned in Section 2.2.3, are dependent on a bandwidth parameter. This parameter is used to control the degree of smoothness in the estimation. Various bandwidth selection methods have been proposed, and for general reviews of bandwidth selection for kernel density and regression, see, for example, Chiu (1996), Heidenreich et al. (2013), and Köhler et al. (2014).

The originally proposed bandwidth selection method was based on minimizing a penalty function (von Davier et al., 2004) in which the continuous score density $f_{h_X}(x)$ deviates as little as possible from the discrete score probabilities r. Recently, other alternative bandwidth selection methods have been suggested to be used within KE, and they are thus included within the GKE framework. The list includes double smoothing (Häggström and Wiberg, 2014), a rule-based method (Andersson and von Davier, 2014), a cross-validation method based on likelihoods (Liang and von Davier, 2014), and traditional leave-one-out cross-validation as well as a hybrid between cross-validation, and the penalty method (Wallin et al., 2021). A particular common feature in these methods is that they are all based on the estimated density, $\hat{f}_{h_X}(x)$. However, as the equating transformation is built using the cumulative distribution functions, it is of interest to explore the possibility of a bandwidth selection method based on CDFs, rather than on densities. A suggestion for such bandwidth selection method within GKE is briefly described in Chapter 6.

2.2.4 The Equating Transformation

In the fourth step of the GKE framework, the equating transformation is built by composing the inverse of the continuized CDF of scores in Y with the continuized CDF of the scores in X (see Eq. (2.3)).

As this composition follows the definition of equating given in Chapter 1 (see also Braun and Holland, 1982), this might be the step in KE that is most difficult to generalize. However, linear versions of the equating transformation (see Eq. (1.3)) can be obtained by conveniently manipulating some of the involved parameters, particularly the bandwidth. Moreover, multiple compositions of CDFs can be used to obtain a *chained* equating transformation (Kolen and Brennan, 2014; von Davier et al., 2004). This type of equating transformation is especially useful when score data are collected under a NEAT design. A further variation of the equating transformation is obtained when conditional score distributions, rather than marginal score distributions, are used to build the equating transformation φ. These types of equating transformations are especially suitable in the context of local equating (van der Linden, 2011). A detailed description of all of these variants of equating transformations that are suitable for use within the GKE framework is provided in Chapter 7.

2.2.5 Evaluating the Equating Transformation

The equating transformation is a functional parameter that has to be estimated using the score data. As such, evaluating the quality of the estimation is an important step in KE. When KE was first proposed, the standard error of equating (SEE) was considered the primary measure to evaluate the accuracy of equating. For this reason, this step was called *calculating the SEE* (von Davier et al., 2004). Because other measures of evaluation have been proposed in the context of KE, we have renamed this step to the more general term *evaluating the equating transformation*. To make Figure 2.1 represents the GKE framework instead of the KE framework, the fifth step should use this new general term, and the SEE equation in the last rectangle in the figure should be replaced with $\hat{\varphi}(x) - \varphi(x)$ to represent a general evaluation of the equating transformation.

We categorize the evaluation measures into two different types of measures: *equating-specific evaluation measures* and *statistical evaluation measures.* The former are those measures that are intended to be used to evaluate equating from a particular feature of either the score distributions or another element involved in the equating process. In the latter, the equating transformation itself is viewed as a functional parameter that can be evaluated as other parameter estimates in statistics are evaluated.

Some examples of equating specific evaluation measures are the percent relative error (PRE), which is used to compare the moments of the observed distribution of Y with those of the distribution of the equated values $\varphi(x)$. The SEE can be categorized as a statistical measure as it is a standard error. In this book we have, however, chosen to categorize it as an equating specific measure, as it is used to measure the uncertainty in the equated values obtained from the equating transformation φ. In the KE framework, this measure is calculated taking into account the type of equating transformation that has been used, the data collection design, and the uncertainty in the estimation of the parameters involved in the presmoothing step. Note that the selection of common items may also affect the SEE (Michaelides and Haertel, 2004, 2014; Lu et al., 2015), and as this has not yet been examined within KE we will not pursue it further in this book. Recently, Marcq and Andersson (2022) proposed a new method for calculating SEE in the EG design that also takes into account variations in the bandwidths. However, they concluded that bandwidth variability effects on the SEE were minimal.

Another useful tool for evaluating the uncertainty in the differences of equated values between two equating transformations is the standard error of equating difference (SEED, von Davier et al., 2004). It is worth mentioning that there are other measures of evaluation not originally developed in the context of KE that can also be used to evaluate it. For instance, a useful tool that has been used in traditional equating is the difference that matters (DTM, Dorans and Feigenbaum, 1994).

The second group of evaluation measures, statistical evaluation measures, is based on the fact that, as with all other statistical estimators, uncertainty exists in the estimation of the equating transformation. To properly evaluate the equating transformation as a functional estimator, standard statistical measures such as bias, standard error (SE), mean squared error (MSE), and root mean squared error (RMSE) can be used (Wiberg and González, 2016; González and Wiberg, 2017), as well as cumulants. Cumulants characterize the shape and structure of probability distributions and can, similar to moments, be used to evaluate how similar distributions are.

The different evaluation measures are discussed in detail in Chapter 8. At this point, we want to emphasize that it is important to use both equating-specific evaluation measures and general statistical evaluation measures when evaluating the equating transformations.

This final step is general and flexible for evaluating the equating transformation in the GKE framework. This means that although we describe a large number of useful measures and methods for evaluating the equating transformation, some of which are most suitable to use with real data and others that are more useful if we conduct a simulation study, more evaluation measures can easily be added to the GKE framework in the future.

2.3 Summary

In this chapter, we reviewed the five steps that define the KE framework. We provided a general overview of each, leaving room to explore multiple methods within each one. Several possibilities for consideration were introduced, with a more in-depth discussion deferred to subsequent chapters.

The GKE framework builds upon the previous KE framework, expanding it in various ways. We anticipate that this broad framework for KE will pave the way for the development of numerous new methods and applications in the field.

Part II

Generalized Kernel
Equating Framework

3

Presmoothing

This chapter demonstrates how to handle sample irregularities that may arise in score distributions. The concept of smoothing aims to mitigate noise in the data, which is assumed to be due to sampling error. While presmoothing can be used as a first step, it is not obligatory in KE if the samples are large enough, although it has proven beneficial in the majority of applications. This chapter primarily centers on three distinct presmoothing modeling approaches: parametric statistical models, nonparametric methods, and methods based on mixture distributions. All of the described methods can be easily incorporated into the GKE framework.

The first method we explore is log-linear modeling, which happens to be the most frequently employed parametric model for presmoothing. The second method is a novel nonparametric proposal that utilizes discrete kernel estimators for presmoothing. Given the discrete nature of score distributions, we have strong confidence in the potential of this approach for presmoothing within the GKE framework.

The third and fourth methods can both be categorized as stemming from mixture distributions. The third method uses item response theory (IRT) modeling, which has been successfully applied in KE applications. Given that the application of IRT to model test and item data is a common practice in many testing programs, IRT presmoothing presents a logical choice.

The fourth method introduces the beta4 model, previously employed to presmooth score distributions in traditional equating methods (Kolen and Brennan, 2014). Its application within KE represents a novel extension and serves as a noteworthy example of the GKE framework's versatility, adaptability, and generalization.

3.1 Presmoothing the Data

Chapter 1 has demonstrated that the linear equating transformation can be estimated using sample statistics (mean and variances) directly obtained from the observed data (see Eq. (1.3)). For this type of equating transformation, the use of sample statistics can be expected to produce accurate equated values. However, the situation differs when the equating transformation is based on the entire score distributions rather than just the first two moments. The upper panel of Figure 3.1 illustrates three hypothetical score distributions for a test comprising 20 items: a symmetrical distribution, a bimodal distribution, and a skewed distribution. The points in the figures correspond to sample relative frequencies for each observed score. It is evident that in all cases, score distributions can exhibit irregularities, potentially resulting in imprecise equating results.

Irregularities due to sampling random error can arise simply because we observe either too few or too many test takers with the same score. Additionally, the frequencies may exhibit nonrandom features such as "teeth" or "gaps" at regular intervals along the score scale (e.g., due to rounding non-integer scores to integers) or "lumps" at zero (resulting from truncation of negative scores) (von Davier et al., 2004). Since the KE transformation is constructed from two score distributions, methods that address these irregularities in the score distributions need to be used. After applying some of these methods, the score distributions are expected to be *smoother*, as demonstrated in the bottom panel of Figure 3.1. Note, smoothing is a property applied to continuous functions. In the equating literature, however, the term *smoothing* has been used to describe the procedure of eliminating irregularities in discrete functions.

As described in Chapter 2, the first step in the KE process is smoothing the score distributions before proceeding with the equating process. This step aims to mitigate the impact of noise in the data, which is assumed to be due to sampling error. The final goal is to obtain equated scores with maximum accuracy. The practice of smoothing when accomplished by fitting statistical models to the score distribution is referred to as *presmoothing*. Although various aspects are to be considered, it is argued that an accurate estimation of the score distribution should preserve essential characteristics of the distributions, such as the moments (Kolen and Brennan, 2014). Note that there is an alternative approach known as *postsmoothing*, in which the equating transformation itself is smoothed, instead of the score distributions. In this book, we will only consider presmoothing methods, as those are used in the first step in the GKE framework.

Research about smoothing in equating includes the comparison of presmoothing vs. postsmoothing methods (e.g., Moses and Liu, 2011), the comparison of log-linear and beta4 models of presmoothing (e.g., Hanson et al., 1994), and studies that combine different methods and the type of smoothing performed (Fairbank, 1987; Hanson, 1991a, 1996; Hanson et al., 1994; Holland and Thayer, 1989; Kolen, 1984, 1991; Kolen and Jarjoura, 1987; Livingston, 1993). A recent discussion about presmoothing in KE with a focus on log-linear modeling and its computational implementation is given in González and Wiberg (2017). The presmoothing methods included in this chapter are log-linear models, discrete kernel estimators, beta4 models, and IRT models. Other presmoothing methods, such as the use of cubic B-splines, are not in-

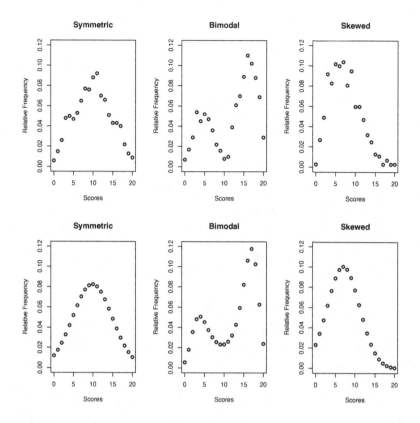

FIGURE 3.1: Three hypothetical score frequency distributions (upper panel) and the results of presmoothing them (bottom panel).

cluded here as these do not preserve the moments of the distribution, and we refer to Cui and Kolen (2009) for more details.

In the next sections, we present these four presmoothing methods and then proceed to discuss the considerations for the model selection within each of them.

3.2   Parametric Statistical Models for Presmoothing

The first class of methods makes use of parametric statistical models that relate the estimated probabilities to model parameters. Estimated score probabilities are then obtained as a function of model parameter estimates. As it was seen in Section 1.2.2, statistical models can be viewed as a collection of probability distributions indexed by parameters and defined on a sample space. For score data, an appropriate probability model would be one for which the probability distribution is defined on a sample space that is a subset of the integer numbers $\mathbb{N}$. Moreover, as each score will be observed with certain probability, the model parameters should be able to capture this feature. A useful probability model that serves to model frequencies occurring with certain probability is the multinomial distribution, which leads to log-linear models.

3.2.1 Polynomial Log-Linear Models

Let N_j be the number of test takers scoring x_j with $p_j = \Pr(X = x_j)$. Assume that $\boldsymbol{N} = (N_1 \ldots, N_J)^t$ follows a multinomial distribution with parameter $\boldsymbol{p} = (p_1, \ldots, p_J)^t$. For test form X under the EG design, we define this parameter as $\boldsymbol{r} = (r_1, \ldots, r_J)^t$. In the log-linear presmoothing method, a log-linear model is fitted to the observed proportions in test form X as

$$\log(r_j) = \beta_0 + \boldsymbol{b}_j^t \boldsymbol{\beta}, \tag{3.1}$$

where β_0 is a parameter acting as a normalization constant, $\boldsymbol{b}_j^t$ is a vector of known constants, and $\boldsymbol{\beta}$ is a vector of parameters. Model (3.1) can also be written in matrix form as

$$\log(\boldsymbol{r}) = \beta_0 + \mathbf{B}^t \boldsymbol{\beta}, \tag{3.2}$$

where $\mathbf{B} = (\boldsymbol{b}_1, \ldots, \boldsymbol{b}_J)$ is a design matrix of known constants. A remarkable property derived from the likelihood equations of this model is the fact that sample and fitted moments are matched, namely,

$$\sum_j b_{ij}(N_j/N_x) = \sum_j b_{ij}\hat{r}_j . \tag{3.3}$$

This result makes the method appealing for presmoothing as the fitted distribution preserves the moments of the observed score distribution. Note that $\sum_j N_j = N_x$. A common choice for the constants is $b_{ij} = x_j^i$, leading to polynomial log-linear models that preserve power moments.[1]

Let T_r denote the highest polynomial degree, $i = 1, \ldots, T_r$. A polynomial log-linear model that can be used to model r_j for test X when a univariate distribution (as in an EG design) is used is

$$\log(r_j) = \beta_0 + \sum_{i=1}^{T_r} \beta_i(x_j)^i = \beta_0 + \beta_1 x_j + \beta_2 x_j^2 + \ldots + \beta_{T_r} x_j^{T_r}. \tag{3.4}$$

---

[1] For alternative types of moments that can be used in this context, see Appendix C in von Davier et al. (2004).

A similar model can be fitted to the score distribution of Y scores by modeling s_k.

Suppose now that we are using score data from a design that produces bivariate distributions (e.g., SG, NEAT, or CB designs). Under this setting, a bivariate log-linear model can easily be used to fit the score data. In addition to main effects, interaction terms can also be added to the model. For example, under the NEAT design, if we let A be the anchor score, we can define $p_{jl} = \Pr(X = x_j, A = a_l)$ and write the model as

$$\log(p_{jl}) = \beta_0 + \sum_{i=1}^{T_r} \beta_i^X (x_j)^i + \sum_{k=1}^{T_a} \beta_k^A (a_l)^k + \sum_{i=1}^{L_r} \sum_{k=1}^{L_a} \beta_{ik}^{XA}(x_j)^i (a_l)^k, \quad (3.5)$$

where T_a denotes the highest polynomial degree in a_l ($k = 1, \ldots, T_a$) and T_r is defined as above. L_r and L_a are the sum limits that serve to accommodate cross moments in the models. The superscripts X, A, and XA are used to differentiate between the β parameters accompanying the powers of the x_j scores, the a_l scores, and the cross products of the test scores $x_j a_l$, respectively.

Note that similar models can be fitted for data collected under the SG and CB designs. For the former, a bivariate log-linear model is fitted to $p_{jk} = \Pr(X = x_j, Y = y_k)$, whereas for the latter, two bivariate log-linear models can be fitted to $p_{(12)jk} = \Pr(X_1 = x_j, Y_2 = y_k)$ and $p_{(21)jk} = \Pr(X_2 = x_j, Y_1 = y_k)$, respectively. Similar models as in (3.4) and (3.5) can be defined for Y under the EG and NEAT designs.

As an example, consider the hypothetical symmetric score distribution shown in Figure 3.1. Figure 3.2 shows the smoothed versions of this score distribution that result from fitting the polynomial log-linear models in Eq. (3.4) for $T_r = 1, 2, 3$. Note that the lines connecting score probabilities are drawn only to facilitate viewing. When only one polynomial degree is used, the smoothing is poor and does not capture the shape of the distribution well. Using either a second or a third polynomial degree yields similar presmoothed distributions, and they are both improvements from using only a single polynomial degree. Plots like this, together with other model evaluation measures, should be used when deciding which presmoothing model to use. Model evaluation and selection of log-linear models for presmoothing are discussed thoroughly in Section 3.5.1.

The use of log-linear models to fit discrete test score distributions is extensively described in, e.g., Rosenbaum and Thayer (1987), Holland and Thayer (1987, 2000), and Moses and Holland (2009). More detailed expositions of the method in the context of KE can be seen in von Davier et al. (2004) or Holland and Thayer (1989). Other studies on different aspects of log-linear presmoothing include, e.g., Livingston (1992), Liu et al. (2009), Moses and Holland (2009), and Moses and Holland (2007). Practical examples are found in Chapter 10 and in González and Wiberg (2017).

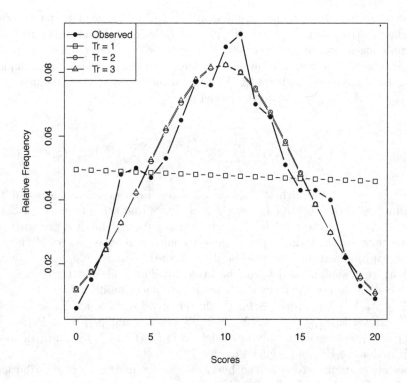

FIGURE 3.2: Observed score frequency distribution and the result of fitting a polynomial log-linear model with power degrees 1, 2, and 3.

3.3 Nonparametric Methods for Presmoothing

Kernel density estimation (Silverman, 1986) is a nonparametric technique used to estimate the density of an unknown continuous probability distribution. Because score data are discrete, an analogue to kernel density estimation that estimates a discrete probability distribution would be a suitable tool for presmoothing. Nonparametric smoothing of discrete variables and functions is an old topic in statistics. For a summary of the literature on this topic see Hall (2001).

In this section, we introduce the use of discrete kernel estimation (e.g., Rajagopalan and Lall, 1995; Kokonendji et al., 2007) to presmooth score distributions. Note that the use of discrete kernel estimation for presmoothing score data was first suggested by González and Wiberg (2024a), who also compared it with presmoothing using log-linear models using empirical data. In this book, we will go further and use the resulting estimated probabilities from these models when performing equating. In the following section, we focus on three discrete associate kernel estimators: the binomial kernel, the discrete triangular kernel, and the Dirac kernel.

3.3.1 Nonparametric Discrete Kernel Estimators

Let $X_1, \ldots, X_N$ be a random sample from a discrete distribution f with probability mass function (PMF) $f(x) = \Pr(X_i = x)$ defined on $\mathcal{X}$. The discrete kernel estimator of f is obtained as

$$\hat{f}(x) = \frac{1}{N} \sum_{i=1}^{N} K_{x,h}(X_i) =: \hat{f}_{N,h,K}(x) , \qquad (3.6)$$

where $h > 0$ is a smoothing parameter and $K_{x,h}(\cdot) = \frac{1}{h} K\left(\frac{x-\cdot}{h}\right)$ is an associate discrete kernel with target x and bandwidth h (Kokonendji et al., 2007). Some choices for K are the binomial, the discrete triangular, and the Dirac kernels.

The binomial kernel estimator defined on $S_x = \{0, 1, \ldots, x+1\}$ can be written as

$$\hat{f}_{N,h,B}(x) := \frac{1}{N} \sum_{i=1}^{N} B_{x,h}(X_i) \qquad (3.7)$$

$$= \frac{1}{N} \sum_{i=1}^{N} \frac{(x+1)!}{X_i!(x+1-X_i)!} \left(\frac{x+h}{x+1}\right)^{X_i} \left(\frac{1-h}{x+1}\right)^{x+1-X_i} 1_{S_x}, \ x \in \mathbb{N} , \qquad (3.8)$$

where $h \in (0, 1]$ is the smoothing parameter, 1_{S_x} is an indicator function, and $B_{x,h}$ is the discrete kernel associated with the binomial distribution with size parameter $x+1$ and probability parameter $\frac{x+h}{x+1}$.

Similarly, the discrete triangular kernel estimator with an arbitrary fixed integer arm a defined on $S_{x,a} = \{x, x \pm 1, \ldots, x \pm a\}$ can be written as

$$\hat{f}_{N,h,T}(x) := \frac{1}{N} \sum_{i=1}^{N} T_{a,h,x}(X_i) \qquad (3.9)$$

$$= \frac{1}{N} \sum_{i=1}^{N} \frac{(a+1)^h - |X_i - x|^h}{(2a+1)(a+1)^h - 2 \sum_{k=0}^{a} k^h} 1_{S_{x,a}}, x \in \mathbb{N} , \qquad (3.10)$$

where $h > 0$ is the smoothing parameter and $T_{a,h,x}$ is the discrete kernel associated with the triangular random variable. Note that the binomial kernel is appropriate for count data with small or moderate sample sizes, while the discrete triangular kernel is recommended for count data with large sample sizes. For categorical data, Kokonendji and Kiessé (2011) proposed a Dirac kernel estimator.

For a fixed number of categories $c \in \{2, 3, ...\}$, define $S_c = \{0, 1, ..., c-1\}$. The Dirac discrete uniform kernel is then defined on S_c as

$$\hat{f}_{N,h,D}(x) := \frac{1}{N} \sum_{i=1}^{N} D_{x,h,c}(X_i) \tag{3.11}$$

$$= \frac{1}{N} \sum_{i=1}^{N} (1-h)1_{\{x\}}(X_i) + \frac{h}{c-1} 1_{S_c \setminus \{x\}}(X_i), \tag{3.12}$$

where $x \in S_c$, $h \in (0, 1]$, and $D_{x,h,c}$ is the discrete kernel associated with the Dirac discrete uniform random variable. Note that if $h = 0$, then the Dirac discrete uniform kernel can be seen as a *naive* kernel estimator in the sense that it estimates exactly the observed relative frequencies, using an associated discrete kernel defined as

$$K_{x,h}(X_i) = \begin{cases} 1, & X_i = x \\ 0, & \text{otherwise} \end{cases} \tag{3.13}$$

for any $x \in \mathcal{X}$ and any $h \geq 0$. Note that in the naive kernel estimator, the bandwidth does not play any role. It simply takes the value 1 if the event $\{X_i = x\}$ occurs, and otherwise it is zero.

Figure 3.3 shows examples of smoothed score distributions when the three described discrete kernels have been used to presmooth a hypothetical symmetric score distribution. In this figure, we can see that the three discrete kernels managed to quite closely capture the observed score data distribution. We provide a more extensive example of using these alternatives to presmooth the test score distribution in Chapter 9.

The estimated probabilities for each test score value can easily be obtained from the discrete kernel estimators. These are then used to obtain the estimated score probabilities, as discussed in Chapter 4.

3.4 Smoothing using Mixture Distributions

Suppose that a probability distribution is indexed by a parameter that itself follows another distribution. This situation is common in latent variable models (e.g., Everitt, 1984), in which a not directly observable variable (a latent variable) is assumed to affect the response variables (i.e., the conditional distribution of responses is indexed by the latent variable). For instance, in IRT models, the items measure a common latent trait typically representing an ability or a psychological attitude.

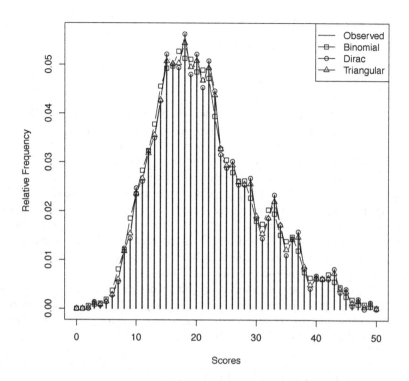

FIGURE 3.3: Observed score frequency distribution and the result of using three discrete kernels.

The distribution that is obtained by marginalizing over the latent variable is called a *mixture* distribution or *compound* distribution (see, e.g., Lindsay, 1995; McLachlan and Peel, 2004; Charalambides, 2005; van der Linden, 2016). Mixtures models are a powerful statistical tool for modeling various types of distributions. In this section, we review two methods based on mixtures that lead to smooth score distributions that are to be used for equating.

3.4.1 Beta4 Models

Lord (1965) developed a model that relates observed and true scores. The discrete mixture distribution that results from mixing the conditional distribution of scores given the true score, with the distribution of true scores, results in a smooth distribution of observed scores. This result applied to presmoothing score distributions, when the true scores are assumed to follow a four-parameter beta distribution, is sometimes referred to as the beta4 method (Kolen and Brennan, 2014). More specifically, let $f(x|\tau)$ be the conditional score distribution given proportion-correct true scores, and let $\psi(\tau)$ be the population distribution of proportion-correct true scores. The observed score distribution can then be obtained as

$$f(x) = \int f(x|\tau)\psi(\tau)d\tau. \tag{3.14}$$

In Lord (1965), $\psi(\tau)$ was assumed to be a four-parameter beta distribution. The conditional distribution of observed scores given the true score was a two-term approximation of the compound binomial distribution. Depending on the value of a parameter, such approximation produces either a binomial or compound binomial conditional distribution. The obtained score distribution $f(x)$ that comes from Eq. (3.14) is referred to as the four-parameter beta compound binomial distribution or the beta4 distribution. Particular cases of the model are the three-parameter beta-binomial model and the well-known beta-binomial model, which corresponds to the case of only two parameters. More details on these distributions can be found in Keats and Lord (1962), Hanson (1991a), and Hanson and Brennan (1990).

From the obtained score distribution $f(x)$ when using a beta4 model, we can get estimated probabilities for each test score value. These are then used to obtain the estimated score probabilities in Chapter 4. An example of smoothing with beta4 models is provided in Chapter 9.

3.4.2   Item Response Theory Models

IRT models (Lord, 1980; Hambleton and Swaminathan, 1985; De Boeck and Wilson, 2004; van der Linden, 2016) are latent variable models for discrete responses obtained from test or questionnaire items intended to measure educational achievement, personality, or attitudes. Fundamental assumptions underlying the IRT models discussed here are i) the unidimensionality of the latent variable, ii) local independence, i.e., the independence of item responses given the latent variable, iii) parameter invariance, i.e., it is assumed that the (items and persons) parameters are similar regardless of which group they are estimated in, and iv) monotonicity, i.e., the probability of a correct response to an item is an increasing monotonic function of the latent trait and the item's parameters. In particular, these IRT models are widely used to score and analyze achievement test data, which are mostly binary-scored. Different versions of IRT models for binary-scored data are obtained depending on the number of item characteristics that are modeled.

For assessments that employ questionnaires or other instruments to measure attitudes or personality traits, items are often scored on a polytomous scale, where multiple response options are available. In these cases, polytomous IRT models like the graded response model (GRM, Samejima, 1969), the partial credit model (PCM, Masters, 1982; Thissen et al., 1995), or the generalized partial credit model (GPCM, Muraki, 1992) can be employed to analyze the responses effectively.

Because IRT models are used in many test-constructing processes, they constitute a convenient choice for equating as well (Skaggs and Lissitz, 1986; Cook and Eignor, 1991; Lord, 1980, Chapter 13). For instance, in IRT OSE (Lord, 1980; Lord and Wingersky, 1984), marginal score distributions are obtained by integrating (or summing) across the ability distribution. The score distributions used in the equipercentile equating function are obtained by cumulating the values of the two corresponding mixtures (see below). It should be noted that IRT OSE can be applied to both binary and polytomously scored items (Kolen and Brennan 2014; Kim et al. 2010).

More formally, let X_{ij} be the random variable representing the answer of test taker i on item j, where $i = 1, \ldots, N_x$; $j = 1, \ldots, n_x$. The response pattern for the test takers can be put into a response matrix of size $N_x \times n_x$, and the observed sum scores can be defined as $X_i = \sum_{j=1}^{J} X_{ij}$. Assuming that test takers with a given ability θ will answer each of the items with probability $p_j(\theta)$ so that $X \mid \theta$ follows a distribution with parameter $p_j(\theta)$, and denoting the distribution of the latent ability θ with $f(\theta)$, the marginal score distribution can be obtained as the following mixture distribution:

$$\Pr(X = x) = \int \Pr(X = x \mid \theta) f(\theta) d\theta . \qquad (3.15)$$

The $\Pr(X = x)$ values are considered to be smoothed IRT-based estimates of the original sample proportions (Holland et al., 2006). These can then be used when obtaining the estimated score probabilities, as shown in the next chapter.

In the psychometric literature, a common choice for the conditional probability distribution $\Pr(X = x \mid \theta)$ has been the compound binomial (Lord and Novick, 1968). The $p_j(\theta)$ parameters can be modeled using one of the common IRT models for either binary or polytomously scored data (e.g., GRM, PCM, GPCM). The distribution $f(\theta)$ is commonly assumed to be a normal distribution. Different ways to solve the integral in Eq. (3.15) are discussed in Kolen and Brennan (2014), p. 199.

3.4.2.1 Binary Item Response Theory Models

Binary-scored items are those for which the test taker receives either a 1 if the item is answered correctly, or 0 if the item is answered incorrectly. In binary-scored IRT models, we model the probability of test taker i, characterized by a person parameter θ_i, answering correctly item j, characterized with a parameter vector ω_j of item features (e.g., item difficulty, item discrimination, and item (pseudo) guessing). Assuming that $\theta_i \sim N(0, \sigma_\theta^2)$, the statistical model becomes

$$(X_{ij} \mid \theta_i, \omega_j) \overset{ind.}{\sim} \text{Bernoulli}(p(\theta_i, \omega_j)), \qquad (3.16)$$

where $p(\theta_i, \omega_j)$ is usually called the item response function which is graphically represented as an item characteristic curve (ICC). A common binary-scored IRT model is the three-parameter logistic model (3PL, Birnbaum, 1968). Let a_j, b_j, and γ_j be the item discrimination, the item difficulty, and the item pseudo guessing parameters, respectively. Then we can define the 3PL model as in (3.16) with

$$p_{ij} = \Pr(X_{ij} = 1 \mid \theta_i, \omega_j) = \gamma_j + (1 - \gamma_j)\frac{\exp[Da_j(\theta_i - b_j)]}{1 + \exp[Da_j(\theta_i - b_j)]}, \qquad (3.17)$$

where D is a scaling constant and $\omega_j=(a_j,b_j,\gamma_j)$. From the 3PL model, we can obtain the two-parameter logistic IRT (2PL) model by setting $\gamma_j = 0$ for all j, and the one-parameter logistic IRT (1PL) model if we also set all $a_j = 1$. Having item parameter estimates[2] so that $\hat{p}_j = p(\theta, \hat{\omega}_j)$, a recursive formula introduced by Lord and Wingersky (1984) can be used to obtain the conditional score distribution for test takers of a given ability, $\Pr(X = x \mid \theta)$. In Chapter 4, we provide a detailed description of the Lord-Wingersky algorithm and more recent alternatives (González et al., 2016) that can be used to obtain the estimated score probabilities.

Presmoothing with binary IRT models in KE was first defined in Andersson and Wiberg (2017) and examples and more details on the computational implementation of IRT KE can be found in Chapter 10 and in Section 7.3 of González and Wiberg (2017).

3.4.2.2 Polytomous Item Response Theory Models

Polytomous IRT models like the GRM and the GPCM can be used to model items that are scored in more than two categories, see e.g., Nering and Ostini (2010), Penfield (2014), and van der Linden (2016) for reviews of available models.

The GRM can be used to model items that are scored in ordered polytomous categories, as when Likert scales are used. The GRM is defined as a cumulative step function, where each step contrasts the probability of obtaining higher scores with the probability of obtaining lower scores on the score scale. The GRM specifies each step function using the 2PL IRT model with a common value for item discrimination for all the steps within an item. Let $p_{j,x}(\theta)$ be the probability of a randomly chosen test taker with ability θ scoring x when there are m_j response categories on item j, and let

$$\Pr(X_j \geq x) = \frac{1}{1 + \exp(-a_j(\theta - b_{j,x}))}, \qquad (3.18)$$

where $b_{j,x}$ and a_j are threshold and item discrimination parameters, respectively.

---

[2]Details on IRT parameter estimation methods can be found in Fischer and Molenaar (1995), Baker and Kim (2004), and Tuerlinckx et al. (2004).

We can then define the GRM as follows:

$$
\begin{aligned}
p_{j,0}(\theta) &= 1 - \Pr(X_j \geq x+1), & x &= 0 \\
p_{j,x}(\theta) &= \Pr(X_j \geq x) - \Pr(X_j \geq x+1), & 1 &\leq x < m_j \\
p_{j,m_j}(\theta) &= \Pr(X_j \geq x), & x &= m_j
\end{aligned}
\tag{3.19}
$$

In the GPCM, we accommodate partial credits of an item with the highest scoring category for item j being m_j. The GPCM models the probability for a randomly chosen test taker with ability θ scoring x on item j as

$$
p_{j,x}(\theta) = \begin{cases} \dfrac{1}{1+\sum_{g=1}^{m_j}\left(\exp(\sum_{v=1}^{g}[a_j(\theta-b_{j,v})])\right)}, & \text{if } x = 0 \\[2ex] \dfrac{\exp(\sum_{v=1}^{m_j} a_j(\theta-b_{j,v}))}{1+\sum_{g=1}^{m_j}\left(\exp(\sum_{v=1}^{g}[a_j(\theta-b_{j,v})])\right)}, & \text{otherwise,} \end{cases}
\tag{3.20}
$$

where the $b_{j,v}$ terms are referred to as step difficulty parameters and $\sum_{v=1}^{m_j}(\theta - b_{j,v}) = 0$.

Presmoothing with polytomous IRT models in KE was first introduced in Andersson (2017) and examples of polytomous IRT KE can be found in Section 7.3 of González and Wiberg (2017) and in Chapter 10 of this book.

3.5 Evaluation of Presmoothing Methods

Kolen and Brennan (2014, p. 69) discuss the importance of the presmoothing models having four characteristics. First, they should give *accurate* estimates of the population distribution, which means that they preserve the moments. Second, they should be *flexible* enough to handle different distributions. Third, there should be a *statistical framework* so it is possible to study the fit. Fourth, the method should improve the estimation, as shown by conducted *empirical research*.

The discussed presmoothing models have all been shown to fulfill all these four criteria. Criteria 1, 2, and 4 can be concluded to be fulfilled from previous studies. For log-linear models, see, for example, Cui and Kolen (2009), Kolen (1984), Kolen (1991), Liu and Kolen (2011), Liu and Kolen (2020), Moses and Holland (2009); for IRT models, see, for example, Lord (1980) and Ogasawara (2003); for discrete kernels, see González and Wiberg (2024a) and González and Wiberg (2024c); and for beta4 models, see Lord (1965), Little and Rubin (1994), and González and Wiberg (2024b). Criterion 3, however, should be examined for each competing models and, in this section, we discuss different strategies to evaluate the fit for each of the described presmoothing methods.

3.5.1   Log-Linear Models

The simplest way to evaluate the fit of a log-linear model used for presmoothing is to graphically inspect how much the observed distribution departs from the smoothed one. The smoothed distributions should not be too different from the observed score distributions.

Because log-linear models are generalized linear models, a more formal alternative way to evaluate model fitting is the use of standard likelihood-based goodness-of-fit statistics. Moses and Holland (2009) recommend using the Akaike information criterion (AIC) (Akaike, 1981). For test X, let $-2\log(\mathscr{L}_{T_r})$ be the log-likelihood ratio chi-square statistic of the model with polynomial degree T_r and u the number of parameters estimated. The AIC criterion can then be calculated as

$$\text{AIC} = -2\log(\mathscr{L}_{T_r}) + 2u. \tag{3.21}$$

It is also possible to use the deviance or the Bayesian information criterion (BIC) (Schwarz, 1978). Letting N_x be the number of observations, BIC can then be calculated as

$$\text{BIC} = -2\log(\mathscr{L}_{T_r}) + \log(N_x)u. \tag{3.22}$$

Note that AIC and BIC are calculated similarly when test Y is used. When comparing models, the model with the smallest values of AIC and BIC is the preferable model.

In von Davier et al. (2004), the authors recommended to start by fitting the simplest univariate log-linear models that only fit the mean and variance of the univariate distribution and then progressively add parameters to fit the higher moments. For the case when univariate distributions are being fitted, the likelihood ratio χ^2 test can be used to select the best model.

Another way to assess model fit is to examine residuals. The Freeman-Tukey residuals (Bishop et al., 1975) are of particular interest as they are approximately standard normally distributed if we can assume that the observed frequencies are Poisson-distributed. Let N_j be the j-th observed score frequency and let $\hat{N}_j$ be the j-th fitted frequency. The Freeman-Tukey residuals are then defined as

$$\text{FT}_j = \sqrt{N_j} + \sqrt{N_j + 1} - \sqrt{4\hat{N}_j + 1}, j = 1, \ldots, J. \tag{3.23}$$

If the observed frequencies only show random variation around the fitted ones, the residuals will exhibit no discernible pattern and will fall within the expected range for nearly normally distributed values, specifically within a range of three standard deviations (Mosteller and Youtz, 1961; Sinharay et al., 2011). For details on how to proceed when examining Freeman-Tukey residuals, please refer to Chapter 4 in González and Wiberg (2017).

It should be noted that the analysis of the Freeman-Tukey residuals is more useful in the univariate case and less useful in the bivariate case because it is possible that there is a large number of zero frequencies in the observed bivariate frequency distribution. An example is therefore given for the EG design in Chapter 9, in which we have used either discrete kernel estimators or a beta4 model in the presmoothing step.

For bivariate distributions, von Davier et al. (2004) recommended fitting first each of the two (marginal) univariate distributions separately, choosing the most parsimonious model for each, and then adding terms of higher order. In this case, the fit indices should be compared for the nested models. Note that for bivariate distributions, the calculated chi-square statistics is not distributed as a χ^2 distribution with the nominal degrees of freedom because observed data might be sparse. Further, Moses and Holland (2010) indicated that it is better to use BIC instead of AIC if bivariate distributions are examined.

Another possibility for evaluating the fit of the bivariate score distributions is to examine the dependencies between the two random variables composing the bivariate score vector (e.g., between X and Y under the SG design, and between X and A and Y and A for the NEAT design). This can be done computing the two sets of (conditional) means, variances, skewness, and kurtoses (e.g., $E(X|Y)$ and $E(Y|X)$, $E(X|A)$ and $E(A|X)$, etc.) of both the fitted and the observed conditional distributions. If the conditional parameters show large discrepancies between the observed and estimated distributions, other potential models should be examined as well. An example of this kind of analysis is shown in Chapter 10.

3.5.2 Discrete Kernel Estimators

Evaluation of discrete kernel estimators follows more or less the same ideas as the ones used for kernel density estimation. Discrete kernels are evaluated in terms of consistency results, and they are normally compared in terms of mean integrated squared error (MISE, Silverman, 1986; Scott, 1992). For a given smoothing parameter h and a given estimator $f_N(x; h)$ of the density $f(x)$, the MISE can be defined as

$$\text{MISE}(h) = E \int (f_N(x; h) - f(x))^2 dx. \tag{3.24}$$

We are looking for models with low MISE values. The same is true for the selection of the smoothing parameter, which is generally obtained by minimizing MISE. The choice of a particular kernel also follows practical considerations according to the data at hand and its characteristics. For instance, Kokonendji and Kiessé (2011) found that a binomial discrete kernel estimator performs better than the empirical estimator for small sample sizes. These authors also investigated the choice of the smoothing parameter according to the classical cross-validation and the more novel excess-of-zeros methods. This is particularly useful for the case in which count data are analyzed and a large proportion of zeros exists in the sample.

In Chapter 9, we illustrate the use of three different discrete kernels in the presmoothing step of the GKE framework. One possible way to evaluate the usage of discrete kernels is to use Freeman-Tukey residuals, as described in Section 3.5.1. An example with discrete kernels is given in Section 9.7.3.

Another possible way is to use chi-square tests or to examine the bias between two discrete distributions. However, further research is required if this is to be a viable alternative. The evaluation of discrete kernels used for presmoothing in equating is a new approach that warrants further research. Issues such as the optimal selection of smoothing parameters so that the estimated probability distribution resembles the observed score frequency distribution, and the choice of a kernel for a particular type of score data, are also topics for future research.

3.5.3   Beta4 Models

To evaluate the fit of different models, it is possible to compare the central moments of the sample data to the ones of the fitted distributions. A key point when using the beta4 model is to ensure that the first four central moments of the fitted distribution agree with those of the sample distribution. This means checking that there are no invalid parameter estimates, such as when the upper limit for the proportion correct true score is above 1. For the case when only three central moments are preserved, Hanson (1991a) developed a strategy to ensure that the first three moments of observed and fitted distributions agree, while the fourth moment of the fitted distribution is made close to the fourth moment of the observed distribution.

It is also possible to use standard statistical methods such as a likelihood ratio test or a standard chi-square goodness-of-fit statistic (Lord, 1965). If we assume that all score points are included when we calculate the chi-square statistic, the degrees of freedom are equal to the number of score points minus 1, minus the number of parameters fit. If J is the number of score points, then the degrees of freedom become $J - 1 - 4 = J - 5$. For more details, see Kolen and Brennan (2014).

One can also use Freeman-Tukey residuals to evaluate the model fit, as described in Section 3.5.1. An example with beta4 models is given in Section 9.7.3.

3.5.4   IRT Models

For the purpose of presmoothing, the optimal IRT model will be the one for which the underlying assumptions are met and which fits the data best. Typical assumptions of IRT models include: (i) unidimensionality, (ii) local independence, (iii) parameter invariance, and (iv) monotonicity.

Unidimensionality means that a test should only measure one dimension. Note that we can still evaluate a multidimensional test with unidimensional IRT models if the items can be divided into the different dimensions and those are examined separately (Wiberg, 2012). Unidimensionality can be checked by performing an exploratory factor analysis, and the assumption is fulfilled if only one clear factor is visible. Concluding that only one dominant factor exists can be done in several ways, and there is not one single method that is preferable. Examples of methods to decide number of factors include using a scree plot (Cattell, 1966), which is a plot of ordered eigenvalues for the factors from the largest to the smallest. In the scree plot, the point where the eigenvalues appear to level off is found. The eigenvalues before this point are assumed to contribute to the explained variance, while the eigenvalues after this point are assumed to contribute relatively little. An example where a scree plot is used to evaluate the assumption of unidimensionality is given in Chapter 10.

Another possibility for examining unidimensionality is to use parallel analysis (Horn, 1965), in which observed eigenvalues of a correlation matrix are compared with those from random data. We can also use the very simple structure (VSS) criterion (Revelle and Rocklin, 1979), which uses a goodness-of-fit test to determine the optimal number of factors to extract. Yet another possibility is to use Wayne Velicer's minimum average partial (MAP) criterion (Velicer, 1976), which is based on the matrix of partial correlations. The average of the squared partial correlation is obtained after extracting one factor at a time. When the minimum average squared partial correlation is reached, no more factors are extracted. Note that VSS and MAP analyses do not necessarily provide recommendations on number of factors. In Chapter 10, we will illustrate the use of the scree plot, parallel analysis, the VSS criterion, and the MAP criterion.

Local independence means that the distributions of item responses are independent within any group of test takers with the same ability θ (Lord and Novick, 1968). Consider n items with responses $\mathbf{x} = (x_1, x_2, \ldots, x_n)$ and define $p_{ij}(\theta)$ as the probability of a test taker i with ability θ answering a binary-scored item j correctly. If local independence holds, then

$$\Pr(\mathbf{X} = \mathbf{x}|\theta) = \prod_{j=1}^{n} p_{ij}(\theta). \tag{3.25}$$

Local independence can be checked by examining the content of the test items to make sure that one item does not reveal the answer to another item. Chi-square tests and different statistics can also be used to examine local independence, as described in Liu and Maydeu-Olivares (2012) or Chen and Thissen (1997). We can also examine the residuals of the used item models, and set a value to indicate which local dependency combinations of items to flag as too high. This approach of using the residuals and an index to flag item pairs for local dependency is described in Yen (1984a) and Chen and Thissen (1997), and is illustrated in Section 10.4.2.1.

Item parameter invariance means that the item parameter estimates should be similar regardless of which groups of persons they are estimated in.

Similarly, person parameter invariance means that the person parameter estimates should be similar regardless of which items are used when they are estimated. This assumption can be examined by estimating the parameters with different groups and comparing the results. For more details on the assumptions of IRT models and for different tools to assess these three assumptions, refer, for example, to Glas and Verhelst (1995), Verhelst (2001), Christensen et al. (2002), Chen and Thissen (1997), Rupp and Zumbo (2006), Liu and Maydeu-Olivares (2012), and references therein.

The assumption of monotonicity means that the probability of a correct response to an item is an increasing monotonic function of the latent ability and the item's parameters. This assumption can be graphically examined by studying an item's ICC to ensure that it exhibits a consistent pattern of increasing probability as the latent ability level increases. An example of examining monotonicity by graphically inspecting ICCs for two items is given in Section 10.4.2.1. It is also possible to examine the estimated item parameters, particularly the item discrimination parameter. For binary IRT models, if monotonicity is fulfilled, the item discrimination parameter should be positive, indicating that higher levels of the latent ability lead to a higher probability of endorsing the item. If the item discrimination parameter is negative or close to zero, it suggests a non-monotonic relationship. Monotonicity can also be examined for ordered polytomous IRT models (Kang et al., 2018). For the GPCM, it means that the category thresholds should be ordered correctly according to the latent trait. This can be examined graphically, as illustrated in Section 10.4.3.1.

Goodness-of-fit tests can be used to examine the fit of IRT models, a topic that has been extensively researched. For a detailed account of goodness-of-fit assessment of IRT models and comparison of different methods, see, for example, Stone and Zhang (2003), and Maydeu-Olivares (2013), and references therein. Recently, Wallin and Wiberg (2024) used the likelihood ratio chi-square, AIC, and BIC measures to evaluate different IRT models in the presmoothing step in KE.

For polytomously scored IRT, it is also common to report goodness-of-fit tests and measures of item and person fit. Examples of item and person fit measures are weighted and unweighted mean squares, as described in Wright and Masters (1982, Chapter 5). These authors proposed the following rules of thumb. Mean-square fit statistics values close to 1.0 are considered good, while values below 1.0 mean that there is an overfit and values above 1.0 mean that there is an underfit. Note that these kinds of rules of thumb always should be evaluated in the specific context they appear. For a test taker i with ability θ who responds to item j, the observed response x_{ij} takes a value between 0 and the maximum possible score of item j. The unweighted mean square referred to as *outfit* is calculated for item j as

$$v_j = \frac{\sum_i^N \left(x_{ij} - \mathrm{E}(X_{ij})/\sqrt{\mathrm{Var}(X_{ij})} \right)^2}{N}, \tag{3.26}$$

where $E(X_{ij})$ is the expected value and $\text{Var}(X_{ij})$ is the variance of x_{ij}. Similarly, for person i we can calculate the unweighted mean squares as follows:

$$v_i = \frac{\sum_j^n \left(x_{ij} - E(X_{ij})/\sqrt{\text{Var}(X_{ij})} \right)^2}{n}. \tag{3.27}$$

As the outfit measures can be sensitive to outliers, it is common to also calculate the weighted mean squares, known as *infit*, i.e., the information weighted fit. Let W_{ij} be a weight defined as the variance of the observed response of person i to item j. Note that the weight is large when the ability of the person matches the difficulty of an item. The infit for item j is then defined as follows:

$$v_j = \frac{\sum_i^N W_{ij} \left(x_{ij} - E(X_{ij})/\sqrt{\text{Var}(X_{ij})} \right)^2}{\sum_j^J W_{ij}}. \tag{3.28}$$

Similarly, for person i, we can calculate the weighted mean squares as follows:

$$v_i = \frac{\sum_j^n W_{ij} \left(x_{ij} - E(X_{ij})/\sqrt{\text{Var}(X_{ij})} \right)^2}{\sum_j^J W_{ij}}. \tag{3.29}$$

We can also use chi-square tests to examine if there is a good item fit. The root mean squared error of approximation (RMSEA) can be used to obtain a magnitude of the item misfit and is defined as

$$\text{RMSEA} = \sqrt{[\max([((\chi^2/df) - 1)/(N - 1)], 0)]}, \tag{3.30}$$

where χ^2 is the chi-square value, *df* is the degree of freedom, and N is the number of test takers. RMSEA decreases as sample size increases for a given χ^2. Hu and Bentler (1999) proposed the following rule of thumb for the interpretation of the RMSEA. When the model fits, the RMSEA has an expected value of 0. Values less than 0.06 are considered a good fit. The standardized maximum likelihood index for each person can also be examined to determine person fit, as described in Drasgow et al. (1985). These authors also proposed the following rule of thumb: if there is a good person fit, the values are close to 0, and values further away than two standard errors are considered a bad person fit. Note these rules of thumb need to be evaluated in the specific context nevertheless. The use of chi-square tests and RMSEA for item fit and standardized maximum likelihood index for person fit are illustrated in Chapter 10.

It should be noted that presmoothing using IRT does not ensure that the presmoothed score probabilities resemble the score frequencies found in the observed data (Holland et al., 2006). However, presmoothing using IRT models has successively been used in IRT KE for both binary and polytomous score data (Andersson and Wiberg, 2014; Andersson et al., 2013; Andersson, 2017;

Andersson and Wiberg, 2017). As many measurement programs use IRT as a modeling tool in their testing processes, kernel IRT OSE will enhance the utility of IRT for the purpose of equating.

3.5.5 Choosing a Presmoothing Model

Wallin and Wiberg (2024) compared the use of IRT models with the use of log-linear models in the KE presmoothing step for binary test data using both real test data and simulated data. Their results indicated that the choice of presmoothing model and which model criteria are used (AIC, BIC, or likelihood ratio chi-square) have an impact on the equating transformation.

The choice of presmoothing model in KE has also been examined for mixed-format tests, i.e., tests which contain both binary and polytomously scored items. Wiberg and González (2021) used real test data in an EG design and concluded that standard errors (SE) were higher when more polytomous items as compared with binary items were used, if a smaller sample size was used and if more discriminating items were included. Recently, comparisons between using log-linear models and IRT models in the presmoothing step when equating mixed format tests have been done both with empirical data (Wallmark et al., 2023b), as well as with simulated data (Wallmark et al., 2023a). The results indicated that if good fit is found using IRT models, then IRT models as compared with log-linear models in the presmoothing step gave smaller SEEs in the lower and upper score scales and also in general smaller root mean squared errors (RMSE).

Recently, González and Wiberg (2024a) and González and Wiberg (2024c) examined the use of discrete kernels in the presmoothing step for binary test data. They discovered that, for instance, the discrete kernels managed to capture the shape of the score distributions better than when log-linear models were used. Further, in Chapter 9, it can be seen that bandwidths (described in Chapter 6) obtained using the observed sample probabilities and those obtained using the binomial discrete kernel are larger than the ones obtained when using beta4 score probabilities when adaptive kernels (described in Section 5.4.5) are used.

3.6 Summary

In this chapter, we have explored four presmoothing methods applicable within the GKE framework. We discussed the evaluation process and how to select an appropriate model within each of these four methods. However, we did not delve into the process of choosing among these four presmoothing methods.

One potential approach is to visually compare them by assessing how closely they align with the observed score distribution. Additionally, we can compare the equating transformations obtained with and without presmoothing the data. If one method yields a curve that more closely resembles the unsmoothed equating transformation compared to the others, it might be considered somewhat superior. This strategy was employed in a study by Kolen and Brennan (2014) to compare log-linear and beta4 models for presmoothing, using plots of the differences between equated scores and raw scores (see Fig. 3.5 in Kolen and Brennan, 2014).

Other approaches involve the use of model fit indices, as seen in Wallin and Wiberg (2024), who examined the choice of presmoothing models (log-linear or IRT) on equated scores. Liu and Kolen (2020) also used various model choice indices to guide their selection of the log-linear model in the presmoothing step.

Nevertheless, we believe that research comparing models from different presmoothing frameworks is an area that merits further exploration. Regardless of the presmoothing model chosen, it is crucial to select a model that fits the data well. Finally, it is important to note that different applications often necessitate distinct presmoothing modeling approaches. Thus we will illustrate different presmoothing approaches in Chapters 9 and 10.

4

Estimating Score Probabilities

In order to estimate the equipercentile equating transformation, we need to know the score cumulative distribution functions for each of the test forms that are to be equated. In the previous chapter, we used statistical models to obtain the estimated probabilities which are used in this chapter as the argument of design functions to get the estimated score probabilities for different data collection designs. The obtained estimated score probabilities will be used in the kernel estimation of the score distributions in a later chapter.

When score data undergo presmoothing, it is important to ensure that the chosen model aligns with the nature of the data, either univariate (EG design) or bivariate (SG, CB, NEAT, and NEC designs) (Livingston and Feryok, 1987). Fitting score distributions is relatively straightforward when dealing with univariate score data. As noted in Chapter 2, regardless of whether univariate or bivariate score data are at hand, we always use univariate or marginal score distribution functions when estimating the equating transformation. This means that careful considerations are needed when dealing with bivariate score data, as we need to obtain the marginal vector of the score probabilities from each of the bivariate score distributions.

This chapter starts with defining design functions, which are functions that map the population score distributions into the vectors of score probabilities. An advantage of using design functions is that all data collection designs can be accommodated in a single and simple way. In this chapter, we will provide descriptions of the design functions applicable to the EG, SG, CB, NEAT, and NEC designs. Next, we will briefly describe the estimated probabilities that are needed when estimating score probabilities.

Besides the use of design functions, we also describe how to obtain the estimated probabilities when other models different from log-linear models have been used in the presmoothing step. In Chapter 3, we described the possibility of utilizing IRT models in the presmoothing step. In this chapter, we elaborate on the methods for obtaining estimated score probabilities using the following approaches: an iterative algorithm described in Lord and Wingersky (1984), the utilization of the Poisson-binomial distribution, and other exact and approximate methods.

4.1   Design Functions

Design functions were introduced in von Davier et al. (2004) as a way to unify all equating designs in the KE framework. They were defined as functions that map the population score distributions into the vectors of score probabilities, r and s.

Let $\mathscr{P}$ be the set of all possible population score distributions arising from different equating designs, and Ω the set of all score probability vectors. A design function, $\mathbf{DF}(\cdot)$, can be formally defined as

$$\mathbf{DF} : \mathscr{P} \mapsto \Omega. \tag{4.1}$$

The elements in $\mathscr{P}$ are denoted by Π and they will be vectors or matrices depending on the considered design. The entries in Π are the probabilities that define the score distributions in the population for each equating design. The set Ω contains the vectors of estimated score probabilities that will be used as weights in the kernel continuization of the score distributions (see Chapter 5). The actual mapping is carried out through a matrix denoted by Δ, which can differ between different designs.

For instance, for the SG design, the observed score data are assumed to be generated by the bivariate score distribution $p_{jk} = \Pr(X = x_j, Y = y_k)$. In this case, Π will be a $J \times K$ matrix with entries p_{jk} $(j = 1, \ldots, J; k = 1, \ldots, K)$, and thus $\mathscr{P} = \{\Pi \in \mathbb{R}^{JK} : p_{jk} > 0; \sum_{j,k} p_{jk} = 1\}$. The Δ matrix should accordingly be one that maps the bivariate p_{jk} into the marginal r_j and s_k $(j = 1, \ldots, J; k = 1, \ldots, K)$. See Section 1.3 for other score distributions in each design.

As in most cases only two score distributions will be needed to obtain the equating transformation, we will often have the case that $\Omega = \Omega_J \times \Omega_K$, where

$$\Omega_J = \{r \in \mathbb{R}^J : r_j > 0; \sum_j r_j = 1\}, \tag{4.2}$$

$$\Omega_K = \{s \in \mathbb{R}^K : s_k > 0; \sum_k s_k = 1\}. \tag{4.3}$$

In cases when more than two score distributions are needed (e.g., for the chained equating transformation used in the NEAT design, as defined later in Section 7.5.1), more than two sets of estimated probabilities will be needed and the set Ω will then be different.

Note that in the SG, EG, and CB designs, the data are always obtained from a common population, and therefore no particular assumptions are needed to accommodate score probabilities defined on a common popula-tion. For all these designs, the population score probabilities only needs to be mapped to the marginal score probabilities. A consequence of this is that the DFs will be linear, having the general form $\mathbf{DF}(\Pi) = \Delta\Pi$. As we will see in the next section, this is not necessarily the case for score probabilities that are not directly defined on a common population, as it is for instance in the NEAT design, where nonlinear DFs will be needed to obtain estimated score probabilities defined on a common population.

In the next section, we give explicit forms for the DF for each design. Our presentation of the DF has been, so far, general in that new versions could be added for other possible data collection designs not considered here by simply varying the appearance of Π and Δ. However, because the formulas for the EG, SG, CB and NEAT designs that follow are directly based on the formulas given in von Davier et al. (2004), where the elements Π were denoted as matrices $\boldsymbol{P}$, $\boldsymbol{Q}$, or vectorized versions of them, we decided to keep such notation as well in the exposition.

4.1.1 EG Design

The DF for the EG design is the simplest among all designs. Because score data come from two independent samples taken from the same population, no particular assumptions are needed to obtain the score probabilities properly defined in a common population. In addition, because the data come from two independent groups, univariate score probabilities $\Pr(X = x_j) = r_j$ and $\Pr(Y = y_k) = s_k$ are readily obtained from the population distribution. This means that, in this case, the DF is just an identity mapping between $\mathscr{P}$ and Ω, as defined below.

Let $\Pi = \boldsymbol{P} \in \mathscr{P}$ with

$$
\boldsymbol{P} = \begin{pmatrix} \Pr(X = x_1) \\ \vdots \\ \Pr(X = x_J) \\ \Pr(Y = y_1) \\ \vdots \\ \Pr(Y = y_K) \end{pmatrix} = \begin{pmatrix} r_1 \\ \vdots \\ r_J \\ s_1 \\ \vdots \\ s_K \end{pmatrix} = \begin{pmatrix} \boldsymbol{r} \\ \boldsymbol{s} \end{pmatrix},
$$

then the DF for the EG design is defined as

$$
\mathbf{DF}(\Pi) = \Delta\boldsymbol{P}, \quad \Delta = \begin{pmatrix} \boldsymbol{I}_J & \boldsymbol{0} \\ \boldsymbol{0}^t & \boldsymbol{I}_K \end{pmatrix}, \tag{4.4}
$$

where $\boldsymbol{I}_J$ is a $J \times J$ identity matrix, $\boldsymbol{I}_K$ is a $K \times K$ identity matrix, and $\boldsymbol{0}$ is a $J \times K$ matrix with all entries 0. It is easily verified that $\mathbf{DF}(\Pi) = \begin{pmatrix} \boldsymbol{r} \\ \boldsymbol{s} \end{pmatrix}$.

4.1.2 SG Design

The SG design produces bivariate score data assumed to be generated by a bivariate score distribution. Let $\Pi = \boldsymbol{P} \in \mathscr{P}$ with

$$
\boldsymbol{P} = \begin{pmatrix} \Pr(X = x_1, Y = y_1) & \cdots & \Pr(X = x_1, Y = y_K) \\ \vdots & \ddots & \vdots \\ \Pr(X = x_J, Y = y_1) & \cdots & \Pr(X = x_J, Y = y_K) \end{pmatrix} = \begin{pmatrix} p_{11} & \cdots & p_{1K} \\ \vdots & \ddots & \vdots \\ p_{J1} & \cdots & p_{JK} \end{pmatrix},
$$

and $\mathrm{vec}(\boldsymbol{P})$ the vectorized version of $\boldsymbol{P}$ such that

$$
\mathrm{vec}(\boldsymbol{P}) = \begin{pmatrix} p_{11} \\ \vdots \\ p_{J1} \\ p_{12} \\ \vdots \\ p_{J2} \\ \vdots \\ \vdots \\ p_{1K} \\ \vdots \\ p_{JK} \end{pmatrix} = \begin{pmatrix} \boldsymbol{p}_1 \\ \vdots \\ \boldsymbol{p}_K \end{pmatrix},
$$

with $\boldsymbol{p}_l = (p_{1l}, p_{2l}, \ldots, p_{Jl})^t$; $l = 1, \ldots, K$. Then the DF for the SG design is defined as

$$
\mathbf{DF}(\Pi) = \Delta \mathrm{vec}(\boldsymbol{P}), \quad \Delta = \begin{pmatrix} \boldsymbol{M} \\ \boldsymbol{N} \end{pmatrix} = \begin{pmatrix} \boldsymbol{1}_K^t \otimes \boldsymbol{I}_J \\ \boldsymbol{I}_K \otimes \boldsymbol{1}_J^t \end{pmatrix} = \begin{pmatrix} \boldsymbol{I}_J & \boldsymbol{I}_J & \cdots & \cdots & \boldsymbol{I}_J \\ \boldsymbol{1}_J^t & \boldsymbol{0}_J^t & \cdots & & \boldsymbol{0}_J^t \\ \vdots & \vdots & \vdots & & \vdots \\ \boldsymbol{0}_J^t & \cdots & \cdots & \boldsymbol{0}_J^t & \boldsymbol{1}_J^t \end{pmatrix},
$$

$$(4.5)$$

where $\boldsymbol{M}$ is a $J \times KJ$ matrix, $\boldsymbol{N}$ is a $K \times KJ$ matrix, $\boldsymbol{1}_J$ and $\boldsymbol{1}_K$ are J-dimensional and K-dimensional vectors of ones, respectively, and $\otimes$ denotes the Kronecker product. It is easily verified that

$$
\mathbf{DF}(\Pi) = \begin{pmatrix} \sum_k p_{1k} \\ \vdots \\ \sum_k p_{Jk} \\ \sum_j p_{j1} \\ \vdots \\ \sum_j p_{jK} \end{pmatrix} = \begin{pmatrix} \boldsymbol{r} \\ \boldsymbol{s} \end{pmatrix}.
$$

4.1.3 CB Design

For the CB design, four different DF can be obtained, depending on the way the data are treated (von Davier et al., 2004). Here, we only show the DF for the case when two independent SG designs are considered.

Let $\Pi = \begin{pmatrix} \boldsymbol{P}_{(12)} \\ \boldsymbol{P}_{(21)} \end{pmatrix} \in \mathscr{P}$, where $\boldsymbol{P}_{(12)}$ is a $J \times K$ matrix with entries $p_{(12)jk} = \Pr(X_1 = x_j, Y_2 = y_k)$, and $\boldsymbol{P}_{(21)}$ is a $J \times K$ matrix with entries $p_{(21)jk} = \Pr(X_2 = x_j, Y_1 = y_k)$. Using the vectorized versions of $\boldsymbol{P}_{(12)}$ and $\boldsymbol{P}_{(21)}$, the DF for the CB design is defined as

$$\mathbf{DF}(\Pi) = \Delta \begin{pmatrix} \mathrm{vec}(\boldsymbol{P}_{(12)}) \\ \mathrm{vec}(\boldsymbol{P}_{(21)}) \end{pmatrix}, \quad \Delta = \begin{pmatrix} w_X \boldsymbol{M} & (1 - w_X)\boldsymbol{M} \\ (1 - w_Y)\boldsymbol{N} & w_Y \boldsymbol{N} \end{pmatrix}, \quad (4.6)$$

where $\boldsymbol{M}$ and $\boldsymbol{N}$ are defined as before, and the weights w_X and w_Y satisfy $0 \leq w_X, w_Y \leq 1$, and are such that

$$r_j = \Pr(X = x_j) = w_X \Pr(X_1 = x_{1j}) + (1 - w_X) \Pr(X_2 = x_{2j}), \quad (4.7)$$
$$s_k = \Pr(Y = y_k) = w_Y \Pr(Y_1 = y_{1k}) + (1 - w_Y) \Pr(Y_2 = y_{2k}). \quad (4.8)$$

It is easily verified that $\mathbf{DF}(\Pi) = \begin{pmatrix} \boldsymbol{r} \\ \boldsymbol{s} \end{pmatrix}$ with $\boldsymbol{r}$ and $\boldsymbol{s}$ having coordinate components as defined in (4.7) and (4.8), respectively.

4.1.4 NEAT Design

As for the SG design, the NEAT design produces bivariate score data. Let $\Pi = \begin{pmatrix} \boldsymbol{P} \\ \boldsymbol{Q} \end{pmatrix} \in \mathscr{P}$, where

$$\boldsymbol{P} = \begin{pmatrix} \Pr(X = x_1, A = a_1) & \cdots & \Pr(X = x_1, A = a_L) \\ \vdots & \ddots & \vdots \\ \Pr(X = x_J, A = a_1) & \cdots & \Pr(X = x_J, A = a_L) \end{pmatrix} = \begin{pmatrix} p_{11} & \cdots & p_{1L} \\ \vdots & \ddots & \vdots \\ p_{J1} & \cdots & p_{JL} \end{pmatrix}$$

and

$$\boldsymbol{Q} = \begin{pmatrix} \Pr(Y = y_1, A = a_1) & \cdots & \Pr(Y = y_1, A = a_L) \\ \vdots & \ddots & \vdots \\ \Pr(Y = y_K, A = a_1) & \cdots & \Pr(Y = y_K, A = a_L) \end{pmatrix} = \begin{pmatrix} q_{11} & \cdots & q_{1L} \\ \vdots & \ddots & \vdots \\ q_{K1} & \cdots & q_{KL} \end{pmatrix}.$$

We will see in Chapter 7 that when data are collected under a NEAT design, two different equating transformations can be defined, leading to either poststratification equating (PSE) or chained equating (CE). Accordingly, the

DFs are different under the NEAT design depending on whether CE or PSE is used.

For NEAT CE (see Section 7.5.1), four different score probability distributions are involved in the estimation of the equating transformation. Thus, we need four different vectors of score probabilities. The DF for NEAT CE is defined as

$$\mathbf{DF}(\Pi) = \Delta \begin{pmatrix} \mathrm{vec}(\boldsymbol{P}) \\ \mathrm{vec}(\boldsymbol{Q}) \end{pmatrix}, \quad \Delta = \begin{pmatrix} \begin{pmatrix} \boldsymbol{M}_P \\ \boldsymbol{N}_P \end{pmatrix} & \mathbf{0} \\ \mathbf{0} & \begin{pmatrix} \boldsymbol{N}_Q \\ \boldsymbol{M}_Q \end{pmatrix} \end{pmatrix}, \quad (4.9)$$

where $\boldsymbol{M}_P$ and $\boldsymbol{M}_Q$ are $J \times JL$ and $K \times KL$ matrices defined as for the CB design, and $\boldsymbol{N}_P$ and $\boldsymbol{N}_Q$ are $L \times JL$ and $L \times KL$ matrices, also defined as for the CB design. It is easily verified that

$$\mathbf{DF}(\Pi) = \begin{pmatrix} \sum_l p_{jl} \\ \sum_j p_{jl} \\ \sum_k q_{kl} \\ \sum_l q_{kl} \end{pmatrix} = \begin{pmatrix} \boldsymbol{r}_P \\ \boldsymbol{t}_P \\ \boldsymbol{t}_Q \\ \boldsymbol{s}_Q \end{pmatrix}, \quad (4.10)$$

where $\boldsymbol{r}_P$ and $\boldsymbol{s}_Q$ are the J-dimensional and K-dimensional vectors of score probabilities defined in populations $\mathcal{P}$ and $\mathcal{Q}$, respectively, and $\boldsymbol{t}_P$ and $\boldsymbol{t}_Q$ are L-dimensional vectors of anchor score probabilities defined in $\mathcal{P}$ and $\mathcal{Q}$, respectively.

All the DFs defined so far map data that are sampled from a common population (see Figure 1.1). For the PSE version of the NEAT design, score probabilities need to be properly defined on a common population. To achieve this goal, we use the concept of a *synthetic population*, given in Braun and Holland (1982) as

$$w\mathcal{P} + (1 - w)\mathcal{Q}, \quad 0 \le w \le 1,$$

and define the following terms. Let $\boldsymbol{r}_P$ and $\boldsymbol{r}_Q$ denote the vectors of score probabilities for X scores in populations $\mathcal{P}$ and $\mathcal{Q}$, respectively. Similarly, let $\boldsymbol{s}_P$ and $\boldsymbol{s}_Q$ be the vectors score probabilities for Y scores in populations $\mathcal{P}$ and $\mathcal{Q}$, respectively. Let t_{Pl} and t_{Ql} be the anchor score probabilities $\Pr(A = a_l)$ defined in populations $\mathcal{P}$ and $\mathcal{Q}$, respectively. Then, using the definition of synthetic population, we define

$$r_j = w r_{Pj} + (1 - w) r_{Qj}, \quad (4.11)$$

$$s_k = w s_{Pk} + (1 - w) s_{Qk}, \quad (4.12)$$

$$t_l = w t_{Pl} + (1 - w) t_{Ql}. \quad (4.13)$$

Note that under the NEAT design, test form X is only administered to population $\mathcal{P}$ and test form Y is only administered to population $\mathcal{Q}$ (see

Section 1.3.6). Thus, the score probabilities r_{Qj} and s_{Pk} cannot be estimated from the collected data. This means that neither r_j nor s_k can be estimated with the data at hand. This is technically an identifiability problem, as noticed by González and San Martín (2018) and San Martín and González (2022). This identifiability problem has traditionally been solved in the equating literature using the conditional score distributions of test scores given the anchor scores, and assuming what is known as the ignorability condition (Rosenbaum and Rubin, 1983) in the following way.

Let $r_P(x_j|a_l)$ and $r_Q(x_j \mid a_l)$ be the conditional score probabilities $\Pr(X = x_j|A = a_l)$ defined in populations $\mathcal{P}$ and $\mathcal{Q}$, respectively. The conditional score probabilities for Y scores, $s_P(y_k|a_l)$ and $s_Q(y_k|a_l)$, can be defined in a similar way. Then, again using the definition of synthetic population, we have

$$r_j = w \sum_l r_P(x_j \mid a_l)t_{Pl} + (1 - w) \sum_l r_Q(x_j \mid a_l)t_{Ql} \qquad (4.14)$$

and

$$s_k = w \sum_l s_P(y_k \mid a_l)t_{Pl} + (1 - w) \sum_l s_Q(y_k \mid a_l)t_{Ql}. \qquad (4.15)$$

Note, however, that r_j and s_k are still not identified because the conditional score probabilities $r_Q(x_j \mid a_l)$ and $s_P(y_k \mid a_l)$ are also not identified. The following identification restrictions (IR) are used to identify r_j and s_k:[1]

IR-PSE1: $r_Q(x_j \mid a_l) = r_P(x_j \mid a_l)$,

IR-PSE2: $s_P(y_k \mid a_l) = s_Q(y_k \mid a_l)$.

Using **IR-PSE1** and **IR-PSE2** in (4.14) and (4.15), the score probabilities become identified, and it is easily verified that they are

$$r_j = \sum_l r_P(x_j \mid a_l)t_l, \qquad (4.16)$$

and

$$s_k = \sum_l s_Q(y_k \mid a_l)t_l, \qquad (4.17)$$

where t_l is defined in (4.13).

The previous results are used to define the DF for NEAT PSE as

$$\mathbf{DF}(\Pi) = \begin{pmatrix} \Delta(t_P, t_Q) \\ \Delta(t_P, t_Q) \end{pmatrix} \begin{pmatrix} \mathrm{vec}(\boldsymbol{P}) \\ \mathrm{vec}(\boldsymbol{Q}) \end{pmatrix}, \qquad (4.18)$$

---

[1]These restrictions were called assumptions **PSE1** and **PSE2** in von Davier et al. (2004).

where

$$\begin{pmatrix} \Delta(t_P, t_Q) \\ \Delta(t_P, t_Q) \end{pmatrix} = \begin{pmatrix} \sum_l \left(w + \frac{(1-w)\sum_k q_{kl}}{\sum_j p_{jl}} \right) \\ \sum_l \left((1-w) + \frac{w \sum_j p_{jl}}{\sum_k q_{kl}} \right) \end{pmatrix}.$$

It is easily verified that

$$\mathbf{DF}(\Pi) = \begin{pmatrix} \sum_l \left(w + \frac{(1-w)\sum_k q_{kl}}{\sum_j p_{jl}} \right) \boldsymbol{p}_l \\ \sum_l \left((1-w) + \frac{w \sum_j p_{jl}}{\sum_k q_{kl}} \right) \boldsymbol{q}_l \end{pmatrix} = \begin{pmatrix} \boldsymbol{r} \\ \boldsymbol{s} \end{pmatrix},$$

where $\boldsymbol{p}_l = (p_{1l}, p_{2l}, \ldots, p_{Jl})^t$ and $\boldsymbol{q}_l = (q_{1l}, q_{2l}, \ldots, q_{Kl})^t$. Note that unlike the previously defined DF, the one for NEAT PSE is a nonlinear function, which is clearly seen from Eq. (4.18).

4.1.5   NEC Design

The NEC design (Wiberg and Bränberg, 2015) is an alternative to the NEAT design for the case when no anchor scores are available. In the NEC design, anchor scores are replaced by the observed frequencies on each level of (a combination of) categorical covariates. Because score data are taken from nonequivalent groups of test takers, the same identifiability problems described for the NEAT PSE arise in this design. The identifiability restrictions imposed are in line with those used in NEAT PSE and replace conditional score distributions of test scores given anchor scores with conditional score distributions of test scores given the values of the covariates.

Let V be a random variable denoting the observed frequency for a combination of covariates with total number of combinations equal to I. For instance, if two covariates with 2 and 3 levels each, respectively, are available, then $I = 2 \times 3 = 6$. Define

$$\boldsymbol{P} = \begin{pmatrix} \Pr(X = x_1, V = v_1) & \cdots & \Pr(X = x_1, V = v_I) \\ \vdots & \ddots & \vdots \\ \Pr(X = x_J, V = v_1) & \cdots & \Pr(X = x_J, V = v_I) \end{pmatrix} = \begin{pmatrix} p_{11} & \cdots & p_{1I} \\ \vdots & \ddots & \vdots \\ p_{J1} & \cdots & p_{JI} \end{pmatrix}$$

and

$$\boldsymbol{Q} = \begin{pmatrix} \Pr(Y = y_1, V = v_1) & \cdots & \Pr(Y = y_1, V = v_I) \\ \vdots & \ddots & \vdots \\ \Pr(Y = y_K, V = v_1) & \cdots & \Pr(Y = y_K, V = v_I) \end{pmatrix} = \begin{pmatrix} q_{11} & \cdots & q_{1I} \\ \vdots & \ddots & \vdots \\ q_{K1} & \cdots & q_{KI} \end{pmatrix}.$$

Then, using similar arguments as for the case of NEAT PSE, the DF for the NEC PSE can be defined as

$$\mathbf{DF}(\Pi) = \begin{pmatrix} \sum_i \left(w + \frac{(1-w)t_{Qi}}{t_{Pi}} \right) \boldsymbol{p}_i \\ \sum_i \left((1-w) + \frac{wt_{Pi}}{t_{Qi}} \right) \boldsymbol{q}_i \end{pmatrix} = \begin{pmatrix} \boldsymbol{r} \\ \boldsymbol{s} \end{pmatrix}, \qquad (4.19)$$

where $\boldsymbol{p}_i$ is the ith column of $\boldsymbol{P}$, $\boldsymbol{q}_i$ is the ith column of $\boldsymbol{Q}$ and the summation is over all the i possible combinations of the values of the covariates.

Instead of using combinations of the covariates, Wallin and Wiberg (2019) proposed using a NEC design with propensity scores. Propensity scores specify the conditional probability of being assigned to a specific treatment (in this case a specific test form) given a covariate vector $\mathbf{D}$ (Rosenbaum and Rubin, 1983). Let ν be equal to 1 if test form Y is administered and 0 if test form X is administered. The propensity score is then defined as $e(\mathbf{D}) = \Pr(\nu = 1|\mathbf{D}) = E(\nu|\mathbf{D})$.

The calculated propensity scores are divided into a number of strata based on the percentiles so that test takers with propensity scores that are placed into the same stratum are assumed to be equivalent with respect to ability. This means that we should choose the number of strata so that the test groups are homogeneous within each stratum based on the covariate distribution. The DF for the NEC PSE with propensity scores are similar to the DF when combinations of covariates are used. However, instead of summing over all possible combinations of the covariate values, we sum over all numbers of stratified propensity scores d as follows:

$$\mathbf{DF}(\Pi) = \begin{pmatrix} \sum_d \left(w + \frac{(1-w)t_{Qd}}{t_{Pd}} \right) \boldsymbol{p}_d \\ \sum_d \left((1-w) + \frac{wt_{Pd}}{t_{Qd}} \right) \boldsymbol{q}_d \end{pmatrix} = \begin{pmatrix} \boldsymbol{r} \\ \boldsymbol{s} \end{pmatrix}. \qquad (4.20)$$

Similarly, Wallin and Wiberg (2019) also proposed using NEC CE with propensity scores instead of using a NEAT CE with anchor test scores. The DF when using propensity scores is defined similarly as when an anchor test form has been used. The difference is that instead of using anchor test scores, we are using the stratified propensity scores. The DF is therefore given here only for completeness.

$$\mathbf{DF}(\Pi) = \Delta \begin{pmatrix} \mathrm{vec}(\boldsymbol{P}) \\ \mathrm{vec}(\boldsymbol{Q}) \end{pmatrix}, \qquad \Delta = \begin{pmatrix} \begin{pmatrix} \boldsymbol{M}_P \\ \boldsymbol{N}_P \end{pmatrix} & \mathbf{0} \\ \mathbf{0} & \begin{pmatrix} \boldsymbol{N}_Q \\ \boldsymbol{M}_Q \end{pmatrix} \end{pmatrix}, \qquad (4.21)$$

where $\boldsymbol{M}_P$ and $\boldsymbol{M}_Q$ are $J \times JL$ and $K \times KL$ matrices defined as for the SG design, and $\boldsymbol{N}_P$ and $\boldsymbol{N}_Q$ are $L \times JL$ and $L \times KL$ matrices, also defined as for the SG design.

4.1.6 Comparison of Designs

Table 4.1 shows the appearance of both $\mathscr{P}$ and Ω, where the DF is defined for each equating design. For the NEAT design, two versions are given (NEAT PSE and NEAT CE), which were explained in Section 4.1.4. Note that we omitted the DF for NEC PSE and NEC CE in this table as they are similar to the DF of the NEAT PSE and NEAT CE designs: see details in Section 4.1.5.

4.2 Estimated Probabilities

Estimated probabilities are those resulting from the fitting of statistical models to score distributions in the presmoothing step. In this book, we have reserved the term *estimated score probabilities* for those that are obtained through the design functions, and which are later used as the weights in the kernel continuization in the third step of the GKE framework. The estimated probabilities can be obtained, as was seen in Chapter 3, from different statistical models, such as log-linear models (Section 3.2.1), discrete kernel estimators (Section 3.3.1), beta4 models (Section 3.4.1), or IRT models (Section 3.4.2).

In the previous section, we saw that the design functions depend on the data collection design. For the EG design, the DF produces an identity mapping, and it is thus straightforward in that case to obtain estimated score probabilities regardless of which statistical model was used to obtain the estimated probabilities. Examples when estimated score probabilities are obtained using discrete kernel estimators under the EG design are given in Section 9.4.2, and with beta4 models in Section 9.4.1. Examples when log-linear models have been used under the EG design can be found in Section 4.3 of González and Wiberg (2017).

Except from the use of log-linear models, for the bivariate case (e.g., NEAT and NEC), it is not as trivial as in the univariate case to obtain estimated probabilities. Examples under the NEAT design using log-linear models are given in Section 10.5.1, and when using IRT models in 10.5.2. How to proceed to obtain estimated score probabilities with IRT models is described in the following section. Obtaining estimated probabilities in the bivariate case when using discrete kernel estimators has not yet been explored, and this is thus a topic for future research. When beta4 models are in use, a proposal for bivariate presmoothing which has not yet been implemented for KE is presented in Hanson (1991b) (see also Lord, 1965).

TABLE 4.1: Elements in $\mathscr{P}$ and the mapped set Ω for each equating design.

Design	$\mathscr{P}$	Ω
EG	$\left\{ \Pi = \binom{r}{s} \in \mathbb{R}^J \times \mathbb{R}^K : r_j > 0; s_k > 0; \sum_j r_j = \sum_k s_k = 1 \right\}$	$\Omega_J \times \Omega_K$
SG	$\left\{ \Pi = P \in \mathbb{R}^{JK} : p_{jk} > 0; \sum_{j,k} p_{jk} = 1 \right\}$	$\Omega_J \times \Omega_K$
CB	$\left\{ \Pi = \binom{P_{(12)}}{P_{(21)}} \in \mathbb{R}^{JK} \times \mathbb{R}^{JK} : p_{(12)jk} > 0; p_{(21)jk} > 0; \sum_{j,k} p_{(12)jk} = \sum_{j,k} p_{(21)jk} = 1 \right\}$	$\Omega_J \times \Omega_K$
NEAT CE	$\left\{ \Pi = \binom{P}{Q} \in \mathbb{R}^{JL} \times \mathbb{R}^{KL} : p_{jl} > 0; p_{kl} > 0; \sum_{j,l} p_{jl} = \sum_{k,l} p_{kl} = 1 \right\}$	$\Omega_J \times \Omega_L \times \Omega_J \times \Omega_L$
NEAT PSE	$\left\{ \Pi = \binom{P}{Q} \in \mathbb{R}^{JL} \times \mathbb{R}^{KL} : p_{jl} > 0; p_{kl} > 0; \sum_{j,l} p_{jl} = \sum_{k,l} p_{kl} = 1 \right\}$	$\Omega_J \times \Omega_K$

4.3   Estimated Probabilities from IRT Models

In Section 3.4.2, we introduced IRT models as potential presmoothing models. Recall that binary IRT models are used to model the probability of correctly answering an item given the person parameter. Polytomously IRT models are instead used when multiple response options are available. The choice of an IRT model depends on the nature of the data, i.e., whether it involves binary or polytomously scored items. After fitting an IRT model, we can obtain the estimated probabilities. In this section, we will describe various methods that make use of the item parameter estimates from an IRT model to estimate the probabilities. We will begin with the most commonly used method, which is an iterative algorithm outlined by Lord and Wingersky (1984).

Alternative methods for estimating probabilities have surfaced in recent years, and we will discuss the utilization of the Poisson's binomial (PB) distribution, followed by an exploration of other approximate and exact methods. While these methods may yield somewhat similar estimated probabilities, they are included here because their applicability can vary depending on specific assumptions and contexts.

This section ends with a brief explanation on how the iterative algorithm can be adapted for use with polytomously scored items and how conditional score probabilities are used in local equating.

4.3.1   The Lord-Wingersky (LW) Algorithm

The most common way to estimate probabilities when we have parameter estimates from an IRT model is to proceed as follows. Using the notation introduced in Section 3.4.2, let X_{ij} be the random variable denoting the answer of test taker i to item j. For the case when the X_{ij} are binary, $X_{ij} = 1$ if the test taker has given a correct answer and $X_{ij} = 0$ if the test taker has given a wrong answer. The conditional distribution of the sum score X for a given ability θ is called the *compound binomial* distribution (Lord and Novick, 1968). Using the compound binomial distribution for the calculation of conditional probabilities is computationally demanding, and that is why recursive methods have been proposed, as described below.

Let $f_l(x \mid \theta)$ be the conditional distribution of sum scores over the first l items for test takers of ability θ. Then, if $p_j(\theta)$ is the probability of answering item j correctly, which is defined from the chosen IRT model, we necessarily have that $f_1(x = 1 \mid \theta) = p_1(\theta)$ and $f_1(x = 0 \mid \theta) = 1 - p_1(\theta)$. The conditional probabilities for $l > 1$ can then be calculated using the following recursion formula given by Lord and Wingersky (1984):

$$
\begin{aligned}
f_l(x \mid \theta) &= f_{l-1}(x \mid \theta)q_l(\theta), & x = 0 & \qquad (4.22) \\
&= f_{l-1}(x \mid \theta)q_l(\theta) + f_{l-1}(x - 1 \mid \theta)p_l(\theta), & 0 < x < l & \qquad (4.23) \\
&= f_{l-1}(x - 1 \mid \theta)p_l(\theta). & x = l, & \qquad (4.24)
\end{aligned}
$$

where $q_l(\theta) = 1 - p_l(\theta)$. For instance, if we have the item difficulty parameter estimates $\hat{b}$ from a 1PL model, the estimated values $p(\theta, \hat{b}_j)$ can be used in the recursion formula above to obtain the conditional observed probabilities for test takers of a given ability θ, $f(x \mid \theta) = \Pr(X = x \mid \theta)$. Finally, by marginalizing over θ, the estimated probabilities $\Pr(X = x)$ can readily be obtained. A hypothetical example showing how the LW algorithm works for a three-item test can be found in Kolen and Brennan (2014) Section 6.6.

For the EG design, the obtained estimated probabilities correspond to the estimated score probabilities r_j from test form X, and s_k can similarly be obtained for test form Y. For the NEAT design, r_j and s_k are obtained differently when IRT models are used in the presmoothing step, and for details we refer to Andersson and Wiberg (2017). A practical example is given in Section 10.5.2.

In the following sections, we briefly describe several alternatives to the Lord-Wingersky algorithm. A detailed description of these alternatives is given in González et al. (2016).

4.3.2 The Poisson's Binomial Distribution

A recently suggested alternative to the Lord-Wingersky algorithm is to use the Poisson's binomial (PB) distribution (González et al., 2016) which is defined in what follows. Let E_n be the random variable representing the total number of successes in n independent Bernoulli trials, where p_j is the probability of success at trial j. It is well known that if $p_j = p$, $\forall j$ (i.e., the identically distributed case) then E_n is Binomial(n, p) distributed. For the case of independent but non-identically distributed Bernoulli trials, E_n instead follows a PB distribution (Wang, 1993; Hong, 2013).

Let $\mathcal{F}_x = \{A : A \subseteq \{1, \ldots, n\}, |A| = x\}$, where $|A|$ denotes the cardinality of the set A. The probability mass function (PMF) of $X = E_n$ is defined as

$$f_n(x; \boldsymbol{p}) = \sum_{A \in \mathcal{F}_x} \left(\prod_{j \in A} p_j \right) \left(\prod_{j \in A^c} (1 - p_j) \right), \qquad (4.25)$$

where A^c denotes the complement of the set A. Because the underlying probability model in IRT is the Bernoulli, and each of the items on a test is independently answered with probability $p_j(\theta)$, it follows immediately that the sum score, X, has the PB distribution which can thus be used to obtain score probabilities.

Let us illustrate how this distribution is used in practice using the following hypothetical example that mimics the one shown in Section 6.6 of Kolen and

Brennan (2014). Assume that $n = 3$ so that we have

$$
\begin{aligned}
\mathcal{F}_0 &= \{\emptyset\} \\
\mathcal{F}_1 &= \{\{1\}, \{2\}, \{3\}\} \\
\mathcal{F}_2 &= \{\{1,2\}, \{1,3\}, \{2,3\}\} \\
\mathcal{F}_3 &= \{\{1,2,3\}\}.
\end{aligned}
$$

Then, the probability of having two successes, $\Pr(X = 2)$ is calculated as

$$
\begin{aligned}
f_n(2, \boldsymbol{p}) &= \left(\prod_{j \in \{1,2\}} p_j \right) \left(\prod_{j \in \{3\}} (1 - p_j) \right) + \left(\prod_{j \in \{1,3\}} p_j \right) \left(\prod_{j \in \{2\}} (1 - p_j) \right) + \\
&\quad \left(\prod_{j \in \{2,3\}} p_j \right) \left(\prod_{j \in \{1\}} (1 - p_j) \right) \\
&= p_1 p_2 (1 - p_3) + p_1 p_3 (1 - p_2) + p_2 p_3 (1 - p_1) \\
&= p_1 p_2 q_3 + p_1 p_3 q_2 + p_2 p_3 q_1.
\end{aligned}
$$

Using similar calculations, it can easily be shown that $\Pr(X = 0) = f_n(0, \boldsymbol{p}) = q_1 q_2 q_3$; $Pr(X = 1) = f_n(1, \boldsymbol{p}) = p_1 q_2 q_3 + p_2 q_1 q_3 + p_3 q_1 q_2$; $\Pr(X = 3) = f_n(3, \boldsymbol{p}) = p_1 p_2 p_3$. It is important to emphasize that these probabilities coincide exactly with those shown in Section 6.6 of Kolen and Brennan (2014), where they were calculated using the Lord-Wingersky algorithm.

4.3.3 Other Approximate and Exact Methods

As an alternative to the direct use of the PB model to obtain estimated probabilities, both approximate and exact alternatives have been proposed in the literature. For example, the normal approximation (NA) approximates the CDF of the PB distribution by

$$
F_X(x) \approx \Phi \left(\frac{x + 0.5 - \mu}{\sigma} \right),
$$

where Φ is the CDF of the standard normal distribution, $\mu = E(X) = \sum_j p_j$, and $\sigma = \sqrt{\mathrm{Var}(X))} = \sqrt{\sum_j p_j (1 - p_j)}$. In psychometrics, this approximation has been used for the compound binomial model (e.g., Lord and Novick, 1968, p. 406).

Volkova (1996) (see also Neammanee, 2005) described an improved version of the normal approximation that corrects for the skewness of the distribution

TABLE 4.2: Score probabilities calculated with different methods.

Score	LW	DFT-CF	NA	RNA	PA
0	0.4430	0.4430	0.3872	0.4285	0.4916
1	0.4167	0.4167	0.4722	0.4267	0.3491
2	0.1277	0.1277	0.1332	0.1257	0.1239
3	0.0126	0.0126	0.0072	0.0186	0.0293

of X. The refined normal approximation (RNA) approximates the CDF of the PB distribution as

$$F_X(x) \approx G\left(\frac{x + 0.5 - \mu}{\sigma}\right),$$

where $G(x) = \phi(x) + \gamma(1 - x^2)\phi(x)/6$, $\gamma = \sigma^{-3}\sum_{j=1}^{n} p_j(1 - p_j)(1 - 2p_j)$, $\phi(x)$ is the density of the standard normal distribution, and μ and σ are defined as before.

Le Cam (1960) (see also Steele, 1994) establishes an inequality that leads to another approximate method for obtaining score probabilities, referred to here as the Poisson approximation (PA). It uses the PMF of the Poisson distribution with parameter μ as defined above, and thus approximates score probabilities as

$$f(x) \approx \frac{\mu^x \exp(-\mu)}{x!}.$$

Finally, we mention an algorithm proposed in Fernandez and Williams (2010) (see also Hong, 2013) in which polynomial interpolation and the discrete Fourier transform (DFT) are used to derive exact closed-form formulas for the PB probability distribution function and CDF. Using this method, the CDF of the PB distribution can be obtained through

$$F_X(x) = \frac{1}{n+1}\sum_{l=0}^{n}\sum_{m=0}^{x} \exp(-\mathrm{i}\omega lm)z_l, \quad x = 0, \ldots, n$$

where $\mathbf{i}$ is the imaginary unit, $\omega = 2\pi/(n+1)$, and z_l is the characteristic function of the PB random variable evaluated at ωl, i.e., $z_l = \prod_{j=1}^{n}[1 - p_j + p_j\exp(-\mathrm{i}\omega l)]$. The method is based on the application of the discrete Fourier transform to the characteristic function (CF) of the PB distribution, and it is accordingly called the DFT-CF method.

Table 4.2 shows a comparison of the different described methods for obtaining score probabilities using the example data from Kolen and Brennan (2014). Note that the score probabilities obtained by the DFT-CF method are identical to those obtained using the LW algorithm. This is due to the fact that both methods are exact and not merely approximations.

It should be mentioned that other related references regarding the distribution of the number of successes in independent binary trials include Walsh

$(1955)^2$, Hoeffding (1956), Darroch (1964), Samuels (1965), Nedelman and Wallenius (1986), and Chapter 6 (Volume 2) in van der Linden (2016). Furthermore, González et al. (2016) showed that the recursive formula proposed by Thomas and Taub (1982) is identical to the LW algorithm.

4.3.4   The LW Algorithm for Polytomous Data

The LW algorithm introduced in Section 4.3.1 is used to obtain the conditional distribution of scores when an IRT model for binary-scored items has been used. When we have items that are polytomously scored, we need to use polytomous IRT models and the LW algorithm needs to be adjusted. Here, we briefly describe an extension of the LW algorithm for the case of polytomously scored items that was first described in Thissen et al. (1995).

Let m_j be the number of response categories of item j and let x_{jk} represent an earned score in category k of item j, $k \in \{1, \dots, m_j\}$. Denote by $f_l(x \mid \theta)$ the conditional distribution of sum scores over the first l items for test takers of ability θ. If $p_{jk}(\theta)$ is the probability of earning a score in category k of item j, then $f_1(x = x_{1k} \mid \theta) = p_{1k}(\theta)$, $k = 1, \dots, m_1$. The conditional probabilities for $l > 1$ can be calculated using the following recursion formula (Kolen and Brennan, 2014)

$$f_l(x \mid \theta) = \sum_{k=1}^{m_j} f_{l-1}(x - x_{jk} \mid \theta) p_{jk}(\theta), \quad \min_l < x < \max_l.$$

where $\min_l$ and $\max_l$ are the minimum and maximum scores after adding the lth item.

The interested reader is referred to Kolen and Brennan (2014) Section 6.10.5. for a hypothetical example showing how this extension of the LW algorithm for polytomous data works in practice.

4.3.5   Conditional Probabilities

Most equating methods are defined to be used with one single transformation that is built on marginal score distributions $F_X(x)$ and $F_Y(y)$. In contrast, local observed-score equating methods (van der Linden, 2011) use conditional score distributions. We define two conditional distributions,

$$F_X(x|\lambda), \quad \lambda \in \Lambda, \tag{4.26}$$

and

$$F_Y(y|\lambda), \quad \lambda \in \Lambda, \tag{4.27}$$

where λ is an index representing individual members in Λ. These two conditional distributions are used to create a family of equating transformations

---

[2]In the psychometric literature, Walsh's method was also used in Lord and Novick (1968, Section 23.10) and later in Yen (1984b).

instead of a single equating transformation. When the parameter λ is unknown, an estimate is needed. In local equating, λ has typically represented the test taker's ability parameter. As ability is seldom directly observable, a proxy for this parameter is needed. For example, in Wiberg and van der Linden (2011) the anchor test score a and the realized observed score $X = x$ were used as a proxy for the ability and as the λ parameter. When anchor test scores are used, the conditional score distributions are defined as $F_X(x|a)$ and $F_Y(y|a)$.

The idea of local equating was originally motivated by Lord's equity principle (Lord, 1980), which claims that observed-score equating (OSE) is only possible when two test forms are perfectly reliable or strictly parallel. In the first local equating proposals, λ was replaced by the ability parameter θ, leading to the so-called local IRT OSE (van der Linden, 2000, 2006).

Later, local IRT observed-score KE (Wiberg, 2016; Andersson and Wiberg, 2017) was proposed as a method to incorporate IRT models within KE and obtain a family of equating transformations. Local IRT observed-score KE has been used under both the EG and the NEAT CE design. Similarly as in local IRT observed-score equating, we let the ability θ be the index of each family member and then the local equating transformation can be defined from the following conditional score distributions:

$$F_{X|\theta}(x; \hat{r}), \tag{4.28}$$

and

$$F_{Y|\theta}(y; \hat{s}). \tag{4.29}$$

The corresponding weights used in the continuization of the score distributions are conditional estimated probabilities given a level of ability. The conditional estimated probabilities can be obtained using the same methods described in the preceding sections, yet omitting the marginalization step.

4.4 Summary

In this chapter, we have described various methods used for estimating score probabilities. A large part of the chapter was devoted to describing the design functions for the different data collection designs. While design functions can be applied directly to sample relative frequencies, they are often applied to presmoothed score distributions. This differs from the LW method and the alternatives described in this chapter, where fitting an IRT model is a necessary step. Besides describing and comparing different methods to obtain estimated probabilities when we have an IRT model, a brief description of the use of conditional probabilities was also included.

One significant advantage of estimating score probabilities based on presmoothed data becomes evident in Chapter 8, when we introduce the concept of the standard error of equating (SEE). SEE relies on the application of the delta method (Rao, 1973), which assumes normality. When score data are presmoothed using log-linear models, the normality assumption is not an issue because the estimates of score probabilities are derived from maximum likelihood estimates. The same holds true for presmoothing methods based on mixtures, such as IRT models and beta4 models. However, for other presmoothing methods that are not likelihood-based (e.g., discrete kernel estimators), it is not immediately guaranteed that estimates of score probabilities will follow a normal distribution. In Chapter 8, we will discuss a method for obtaining SEE that does not rely on normality assumptions. Such a method can be a valuable alternative to the SEE obtained with the delta method, which relies on normality assumptions.

5

Continuization

In Chapter 2, we established that using continuization is a necessary step to transform discrete score distributions into continuous versions. In traditional equating, linear interpolation has often been employed for continuization. An advantage of linear interpolation is that it is fast, easy to understand, and does not require any sophisticated software. A disadvantage is that it is quite crude and does not necessarily give a very smooth curve. Other continuization methods also exist, such as exponential families, and the continuized log-linear method.

In KE, kernel density estimation has been the prevalent method for continuization. In this chapter, we begin by revisiting the theory of kernel density estimation. Kernel density estimation provides a more flexible continuization than linear interpolation and offers several options. Depending on how parts of the continuization are varied, either traditional equipercentile-like equating, linear equating, or something in between can be obtained.

Next, we introduce an alternative approach using the convolution of distribution functions. Convolution is a mathematical operation that enables us to determine the distribution function of a sum of two random variables by utilizing the distributions of the individual variables being summed. Convolutions are formally defined in Section 5.2, as they have not been previously employed in KE. Because the use of convolutions results in both kernel density and distribution estimators of score distributions, we term it *kernel continuization* to distinguish it from other continuization techniques. We demonstrate that kernel density estimation can be seen as the outcome of convolving two probability distribution functions. A general strategy is developed in the context of kernel continuization in which score distribution functions are convolved with a continuous probability distribution function. In particular, the convolutions are used to justify the use of different kernel continuization alternatives within GKE.

In the subsequent section, we describe kernel density estimation within GKE and outline specific cases of continuous distributions that lead to various kernel continuization alternatives. The chapter ends with a description of the possible use of cumulants within GKE.

5.1 Kernel Density and Distribution Estimation

Kernel smoothing techniques belong to a broader category of methods used for nonparametric function estimation (Rao, 1983; Hart, 1997). When the objective is to estimate probability density functions, a typical choice for the kernel is a unimodal probability density function that is symmetric around 0. This selection ensures that the estimator itself represents a probability density, which is essential when dealing with equating as it involves probability distribution functions. Kernel density estimation is a widely used approach for estimating both density and distribution probability functions (Silverman, 1986; Wand and Jones, 1995; Scott, 2015).

Let $X_1, \ldots, X_N$ be a sample from an unknown distribution with density f. The kernel density estimator of f is defined as

$$\hat{f}(x) = \frac{1}{Nh} \sum_{i=1}^{N} k\left(\frac{x - X_i}{h}\right), \tag{5.1}$$

where k is a function such that

$$\int k(t)dt = 1, \quad \int tk(t)dt = 0, \quad \int t^2 k(t)dt < \infty, \tag{5.2}$$

and h is a smoothing parameter called the bandwidth. An estimator of the cumulative distribution function is easily obtained by integrating (5.1) as

$$\hat{F}(x) = \int_{-\infty}^{x} \hat{f}(t)dt = \frac{1}{N} \sum_{i=1}^{N} K\left(\frac{x - X_i}{h}\right), \tag{5.3}$$

where $\int_{-\infty}^{x} k(t)dt = K(x)$.

Nonparametric kernel density and distribution estimation are well-suited for continuous data. However, score distributions are primarily discrete, and it is therefore essential to employ appropriate methods for estimating discrete probability distributions. For instance, one could use the empirical distribution function or a more advanced kernel estimator tailored for discrete distributions (as discussed in Chapter 3). While these methods may offer better coherence for estimating discrete probability distributions, the resulting estimated distributions will remain discrete. Consequently, defining the equating transformation as in Eq. (1.2) becomes impractical (Varas et al., 2019, 2020).

Our proposed strategy for obtaining continuous versions of score distributions in KE, referred to as *continuization* in KE, is to consider the *convolution* of the discrete score distribution function with the distribution function of a continuous random variable. In the subsequent section, we provide a formal definition of the concept of convolution.

5.2 Convolutions

Convolution is a mathematical operation applied to functions, resulting in the generation of another function (e.g., Sundt and Vernic, 2009; Casella and Berger, 2002). When these functions represent probability distributions, the convolution operation enables us to determine the distribution of the sum of two independent random variables. If $X \sim f_X$ and $Y \sim f_Y$ are two independent random variables, then the distribution f_Z of $Z = X + Y$ is obtained as

$$f_Z(z) = (f_X * f_Y)$$
$$= \begin{cases} \int_{-\infty}^{\infty} f_X(w) f_Y(z - w) dw, & \text{if } X \text{ and } Y \text{ are continuous;} \\ \sum_w p_X(w) p_Y(z - w), & \text{if } X \text{ and } Y \text{ are discrete; and} \\ \sum_w p_X(w) f_Y(z - w), & \text{if } X \text{ is discrete and } Y \text{ is continuous,} \end{cases}$$
$$(5.4)$$

where the $*$ sign denotes the convolution operation.

As an example, consider the convolution of the empirical probability density function (EPDF) with a continuous distribution k. The EPDF is defined as

$$f_N(x) = \frac{d}{dx} F_N(x) = \frac{1}{N} \sum_{i=1}^{N} \delta(x - x_i), \qquad (5.5)$$

where $F_N(x)$ is the usual empirical distribution function, and $\delta(\cdot)$ is the Dirac delta function (Scott, 2015). Note that $f_N(x)$ is a uniform discrete distribution that assigns probability $1/N$ for each data point. The distribution of $Z = X + hY$, when X and Y are independent discrete and continuous random values $X \sim f_N$, $Y \sim k$, respectively, and $h > 0$, can be obtained using convolution. Interestingly, it can be shown to be equivalent to the kernel density estimator in Eq. (5.1). In fact, if we define a random variable $W = hY$, then the density of this random variable is given by

$$f_W(w) = \frac{1}{h} k \left(\frac{w}{h} \right), \qquad (5.6)$$

and it follows that the distribution of $Z = X + W$ can be obtained as the convolution of (5.5) and (5.6) in the following way:

$$f_Z(z) = \int_R f_X(z - w) \frac{1}{h} k \left(\frac{w}{h} \right) dw \qquad (5.7)$$

$$= \int_R \frac{1}{N} \sum_{i=1}^{N} \delta(z - w - x_i) \frac{1}{h} k \left(\frac{w}{h} \right) dw \qquad (5.8)$$

$$= \frac{1}{N} \sum_{i=1}^{N} \int_R \delta(z - w - x_i) \frac{1}{h} k\left(\frac{w}{h}\right) dw \qquad (5.9)$$

$$= \frac{1}{N} \sum_{i=1}^{N} \frac{1}{h} k\left(\frac{z - x_i}{h}\right) \qquad (5.10)$$

$$= \frac{1}{Nh} \sum_{i=1}^{N} k\left(\frac{z - x_i}{h}\right), \qquad (5.11)$$

considering that $\delta(x) = \delta(-x)$ and $\int_{-\infty}^{+\infty} \delta(x - a) f(x) dx = f(a)$.

When obtained as the convolution of a discrete and a continuous distribution, a key feature of f_Z is that it results in a continuous distribution of Z (see Theorem 3.3.2 in Lukacs, 1970). This result is used to obtain continuous approximations of the discrete score distributions so that the KE function defined in (2.3) (see also Chapter 7) can be properly computed.

Note that because the estimator f_Z is itself a density, characteristics of the distribution, such as its moments, can easily be obtained. For instance, the mean and variance of Z are obtained as

$$\mathrm{E}(Z) = \int z f(z) dz = \frac{1}{Nh} \sum_{i=1}^{N} \int zk\left(\frac{z - X_i}{h}\right) dz \qquad (5.12)$$

$$= \frac{1}{N} \sum_{i=1}^{N} \int (X_i + uh) k(u) du \qquad (5.13)$$

$$= \frac{1}{N} \sum_{i=1}^{N} \left[X_i \int k(u) du + h \int uk(u) du \right] \qquad (5.14)$$

$$= \frac{1}{N} \sum_{i=1}^{N} X_i, \qquad (5.15)$$

and similar calculations show that

$$\mathrm{E}(Z^2) = \int z^2 f(z) dz = \frac{1}{N} \sum_{i=1}^{N} X_i^2 + h^2 \kappa, \qquad (5.16)$$

where $\kappa = \int u^2 k(u) du$, so that the variance of Z becomes $\mathrm{Var}(X) + h^2\kappa$. Note that while $\mathrm{E}(Z) = \mathrm{E}(X)$, $\mathrm{Var}(Z) > \mathrm{Var}(X)$. Also note that these results are straightforwardly obtained by direct application of expectation and variance operators on $Z = X + hY$.

In the following section, we propose a formalization of the continuization step in KE showing how the convolution technique is used to obtain continuous approximations of the discrete score distribution functions.

5.3   General Kernel Continuization

Below, we describe continuization for the X score random variable, although the definitions for Y are analogous. Let V be a continuous random variable and let $X(h_X)$ be a continuized[1] score random variable defined as

$$X(h_X) = X + h_X V. \tag{5.17}$$

From (5.17), it can be seen that a continuized score is the sum of the raw score and a continuous random variable weighted by a bandwidth parameter. From the previous section, we know that the expected value of X is preserved by $X(h_X)$, but not its variance. It is, however, straightforward to linearly transform Eq. (5.17) so that the variance is also preserved. As a matter of fact, define

$$X(h_X) = a_X(X + h_X V) + (1 - a_X)\mu_X , \tag{5.18}$$

where

$$a_X = \sqrt{\frac{\sigma_X^2}{\sigma_X^2 + \sigma_V^2 h_X^2}}. \tag{5.19}$$

It then follows that $E(X(h_X)) = \mu_X$ and $Var(X(h_X)) = \sigma_X^2$; i.e., the first two moments of X are preserved in $X(h_X)$. Here, $\mu_X = \sum_j x_j r_j$, $\sigma_X^2 = \sum_j (x_j - \mu_X)^2 r_j$, and V is a continuous random variable with distribution k with mean 0 and variance σ_V^2.

Using the results of the previous section, Theorem 1 states that an explicit expression for the density of $X(h_X)$ can be obtained via convolution.

Theorem 1. *Let $X(h_X)$ be a continuized random variable defined as in (5.18), and define the random variables $T = a_X X$ and $W = a_X h_X V + (1 - a_X)\mu_X$. Then, the density function of $X(h_X)$ is obtained by convolving the distributions of T and W and is shown to be*

$$f_{h_X}(x) = \sum_j r_j k(R_{jX}(x)) \frac{1}{a_X h_X} , \tag{5.20}$$

where $R_{jX}(x) = \frac{x - a_X x_j - (1 - a_X)\mu_X}{a_X h_X}$, $r_j = Pr(X = x_j)$, and $V \sim k$.

Proof. For the random variables $T = a_X X$ and $W = a_X h_X V + (1 - a_X)\mu_X$, use the transformation of variables theorem to obtain the distributions of T and W and then apply the formula in (5.4) to obtain the distribution of $X(h_X)$ □.

---

[1]A random variable that results from adding a continuous random variable to a discrete random variable has also been called a *continued* random variable by other authors (Denuit and Lambert, 2005).

The CDF of $X(h_X)$ can be obtained by integrating (5.20) as

$$F_{h_X}(x) = \int_{-\infty}^{x} f_{h_X}(t)dt = \sum_j r_j K(R_{jX}(x)) , \qquad (5.21)$$

where $K(v) = \int_{-\infty}^{v} k(t)dt$ is the CDF of V.

Alternative derivations of these results appeared in von Davier et al. (2004). These authors had another approach as they first derived the CDF of $X(h_X)$ as follows:

$$\begin{aligned}
F_{h_X}(x) &= \Pr(X(h_X) \leq x) && (5.22) \\
&= \Pr(a_X(X + h_X V) + (1 - a_X)\mu_X \leq x) && (5.23) \\
&= \Pr(a_X h_X V \leq x - a_X X - (1 - a_X)\mu_X) && (5.24) \\
&= \sum_j \Pr(a_X h_X V \leq x - a_X x_j - (1 - a_X)\mu_X \mid X = x_j) \Pr(X = x_j) && \\
& && (5.25) \\
&= \sum_j \Pr\left(V \leq \frac{x - a_X x_j - (1 - a_X)\mu_X}{a_X h_X}\right) r_j && (5.26) \\
&= \sum_j r_j F_V(R_{jX}(x)) && (5.27) \\
&= \sum_j r_j K(R_{jX}(x)) . && (5.28)
\end{aligned}$$

This result appears as Theorem 4.1 in their book. The density function $f_{h_X}(x)$ can then be found by differentiating $F_{h_X}(x)$ in x, which yields (5.20).

Although $K = \Phi$, the standard normal (or Gaussian) distribution function is a common choice of kernel, in the following section we show that other alternative kernels can also be used for equating. This comes from the fact that K can in principle be chosen to be any unimodal probability distribution function symmetric around 0. This last condition ensures that F_{h_X} is also a probability distribution, which is a fundamental requirement to obtain the KE transformation.

5.4   Different Kernels

As will be shown in Chapter 7, the KE transformation is computed as the functional composition of the continuous distribution functions $F_{h_Y}^{-1}(y)$ and $F_{h_X}(x)$, where

$$F_{h_X}(x) = \sum_j r_j K(R_{jX}(x)), \text{and} \qquad (5.29)$$

$$F_{h_Y}(y) = \sum_k s_k K(R_{kY}(y)). \qquad (5.30)$$

As seen in the previous section, the kernel K is associated with the probability distribution of the continuous random variable V used in (5.18). There are different choices for the distribution of V. For instance, von Davier et al. (2004) considered the Gaussian distribution, Holland and Thayer (1989) proposed to use the uniform distribution, and Lee and von Davier (2011) suggested the use of the uniform and logistic distributions. Another possibility is the Epanechnikov (Epanechnikov, 1969) distribution, which was proposed to be used for KE by Cid and von Davier (2015) and González and von Davier (2017). Adaptive kernels have also been proposed in the past (Cid and von Davier, 2015; González and von Davier, 2017). In what follows, we summarize different alternatives for the distribution of V used for kernel continuization.

5.4.1 Gaussian Kernel

The Gaussian CDF for the random variable $V \sim N(0, \sigma_V^2)$ is defined as

$$F_V(v) = \int_{-\infty}^{v} \frac{1}{\sqrt{2\pi\sigma_V^2}} \exp\left\{ -\frac{1}{2\sigma_V^2} t^2 \right\} dt , \qquad (5.31)$$

and the density function is defined as

$$f_V(v) = \frac{1}{\sqrt{2\pi\sigma_V^2}} \exp\left\{ -\frac{1}{2\sigma_V^2} v^2 \right\}. \qquad (5.32)$$

For this choice of the distribution, the equating transformation is computed using the continuized score distributions $F_{h_X}(x)$ and $F_{h_Y}(y)$ defined in (5.29) and (5.30) with $K = F_V = \Phi$. Note that, σ_V^2 may take different values but a common choice is 1, which was used in von Davier et al. (2004). The density functions of $X(h_X)$ and $Y(h_Y)$ are defined as in (5.20) with $k = f_V = \phi$. For more details about the use of the Gaussian kernel within kernel equating refer to von Davier et al. (2004).

5.4.2 Logistic Kernel

The logistic CDF for the random variable V is defined as

$$F_V(v) = \frac{1}{1 + \exp(-v/s)}, \qquad (5.33)$$

where s is a scale parameter and V has mean $E(V) = 0$ and variance $\sigma_V^2 = \pi^2 s^2/3$. The corresponding density is defined as

$$f_V(v) = \frac{\exp(-v/s)}{s(1 + \exp(-v/s))^2}. \tag{5.34}$$

The standard logistic distribution is obtained by setting $s = 1$ and it has $\text{Var}(V) = \pi^2/3$. A distribution with mean 0 and unit variance can be obtained for $s = \sqrt{3}/\pi$ and is called the rescaled logistic distribution. For more details about the use of the logistic kernel within kernel equating refer to Lee and von Davier (2011).

5.4.3   Uniform Kernel

If V is a uniform continuous random variable in the interval $[-\epsilon, \epsilon]$ for some $\epsilon > 0$, then the CDF of V is given by

$$F(v) \;\; = \;\; \begin{cases} 0 & v < -\epsilon \\ (v + \epsilon)/2\epsilon & -\epsilon \leq v < \epsilon \\ 1 & v \geq \epsilon \end{cases} \tag{5.35}$$

and the corresponding density function of V is

$$f(v) \;\; = \;\; \begin{cases} 1/2\epsilon & -\epsilon \leq v \leq \epsilon, \\ 0 & \text{otherwise.} \end{cases} \tag{5.36}$$

The mean for this random variable is 0, and the variance is $\epsilon^2/3$. When $\epsilon = 1/2$, this distribution is often known as a *standard uniform* with $\text{Var}(V) = 1/12$. When V is rescaled to have identity variance (i.e., $\epsilon = \sqrt{3}$), the distribution is called *rescaled uniform*.

Replacement of (5.35) in (5.29) and (5.30) is not as straightforward as for the previous described distributions. It is, however, possible to show that when V is uniformly distributed, then the continuized random variable $X(h_X)$ has the CDF given by

$$F_{h_X}(x) \;\; = \;\; \sum_{j \in A} r_j + \sum_{j \in B} \left\{ r_j \frac{R_{jX}(x) + \epsilon}{2\epsilon} \right\}, \tag{5.37}$$

where $A = \{j : R_{jX}(x) \geq \epsilon\}$ and $B = \{j : -\epsilon \leq R_{jX} \leq \epsilon\}$. The corresponding density function of $X(h_X)$ is

$$f_{X_{h_X}}(x) \;\; = \;\; \frac{1}{a_X h_X} \sum_{j \in B} \frac{r_j}{2\epsilon}. \tag{5.38}$$

For more details about the use of the uniform kernel within kernel equating refer to Holland and Thayer (1989) and Lee and von Davier (2011).

5.4.4 Epanechnikov Kernel

The Epanechnikov kernel (Epanechnikov, 1969) is defined for a random variable V with CDF given by

$$
F(v) = \begin{cases} 0 & v < -1 \\[2mm] \dfrac{3v - v^3 + 2}{4} & -1 \le v \le 1 \\[2mm] 1 & v > 1 \end{cases} ,
$$

and the corresponding density function

$$
f(v) = \frac{3}{4}(1 - v^2)1_{|v| \le 1} . \tag{5.39}
$$

For this random variable, it is easily verified that $E(V) = 0$ and $Var(V) = \frac{1}{5}$. Using (5.29), the Epanechnikov kernel continuization for $X(h_X)$ becomes

$$
F_{h_X}(x) = \sum_{\mathcal{J}} \frac{r_j \left(3R_{jX}(x) - R_{jX}^3(x) + 2\right)}{4} + \sum_{\mathcal{K}} r_j , \tag{5.40}
$$

where $\mathcal{J} = \{j : -1 \le R_{jX} \le 1\}$ and $\mathcal{K} = \{j : R_{jX} > 1\}$. For more details about the use of the Epanechnikov kernel within kernel equating refer to Cid and von Davier (2015) and González and von Davier (2017).

5.4.5 Adaptive Kernels

Adaptive kernels (e.g., Silverman, 1986; Terrell and Scott, 1992) allow the bandwidth parameter to vary across the data points in the score distribution. This means that in the general definition of the kernel density estimator given in (5.1), the fixed value h is replaced by h_i $(i = 1, \ldots, N)$, leading to

$$
\hat{f}(x) = \frac{1}{N} \sum_{i=1}^{N} \frac{1}{h_i} k\left(\frac{x - X_i}{h_i}\right) . \tag{5.41}
$$

Adaptive kernels were first introduced within KE by Cid and von Davier (2015) followed by González and von Davier (2017). Here, we introduce it slightly differently using the theory developed in the previous sections. Define the continuized score random variable as

$$
X(h_{jX}) = a_{jX}(X + h_{jX}V) + (1 - a_{jX})\mu_X , \tag{5.42}
$$

with

$$
a_{jX} = \sqrt{\frac{\sigma_X^2}{\sigma_X^2 + \sigma_V^2 h_{jX}^2}} , \tag{5.43}
$$

$$h_{jX} = \lambda_j h_X, \tag{5.44}$$

where h_X is a fixed bandwidth parameter, and λ_j is a local bandwidth factor to be defined later. For this definition of the continuized score random variable, it is easy to check that $E(X(h_{jX})) = \mu_X$ and $Var(X(h_{jX})) = \sigma_X^2$; i.e., the first two moments of X are preserved in $X(h_{jX})$.

Similarly to the case of the continuized score random variable when a fixed bandwidth parameter is used for the derivation of its density, Theorem 2 shows that an explicit expression for the density of $X(h_{jX})$ can be obtained via convolution.

Theorem 2. *Let $X(h_{jX})$ be a continuized random variable defined as in (5.42), and define the random variables $T = a_{jX}X$ and $W = a_{jX}h_{jX}V + (1 - a_{jX})\mu_X$. Then, the density function of $X(h_{jX})$ is obtained by convolving the distributions of T and W and is shown to be*

$$f_{h_{jX}}(x) = \sum_j r_j k(R_{jX}(x)) \frac{1}{a_{jX}h_{jX}}, \tag{5.45}$$

where $R_{jX}(x) = \frac{x - a_{jX}x_j - (1 - a_{jX})\mu_X}{a_{jX}h_{jX}}$, $r_j = \Pr(X = x_j)$, and $V \sim k$.

Proof. The proof is similar to the one for Theorem 1 by defining $T = a_{jX}X$ and $W = a_{jX}h_{jX}V + (1 - a_{jX})\mu_X$ $\square$.

The CDF of $X(h_{jX})$ can be obtained by integrating (5.45) to get

$$F_{h_{jX}}(x) = \int_{-\infty}^{x} f_{h_{jX}}(t)dt = \sum_j r_j K(R_{jX}(x)). \tag{5.46}$$

Note that the same results can be obtained using Equations (5.22)–(5.28) or Theorem 4.1 in von Davier et al. (2004). The adaptive kernel continuization thus has the following general form:

$$F_{h_{jX}}(x) = \sum_j r_j K\left(\frac{x - a_{jX}x_j - (1 - a_{jX})\mu_x}{a_{jX}h_{jX}}\right). \tag{5.47}$$

We now describe the details on how to select the λ_j parameter, and for this, we consider a Gaussian adaptive kernel so that $K(\cdot) = \Phi(\cdot)$. Silverman (1986) suggested the following steps to obtain λ_j:

i. Find a pilot estimate of the density, $\widetilde{f}(t)$, such that $\widetilde{f}(X_j) > 0 \; \forall j$.

ii. Define a local bandwidth factor λ_j as

$$\lambda_j = \left(\frac{\widetilde{f}(X_j)}{\bar{g}}\right)^{-\alpha},$$

where $\bar{g}$ is the geometric mean of $\widetilde{f}(X_j)$ and α is a sensitivity parameter satisfying $0 \le \alpha \le 1$. Silverman's recommendation is to use $\alpha = 0.5$.

To obtain λ_j, we propose as a pilot estimate

$$\widetilde{f}(x) = \sum_j r_j \phi \left(\frac{x - a_X x_j - (1 - a_X)\mu_x}{a_X h_X} \right) \frac{1}{a_X h_X}, \tag{5.48}$$

where h_X can be obtained using any of the bandwidth selection methods to be described in Chapter 6.

Following the strategy described above, we can also obtain $F_{h_{jY}}$, such that the adaptive KE transformation becomes $\varphi(x) = F_{h_{jY}}^{-1}(F_{h_{jX}}(x))$.

5.4.6 The Percentile Rank Method

The percentile rank method (PRM) originated as a graphical method for equating scores (e.g., Angoff, 1971). This method, also known as equipercentile equating, determines equated scores by identifying values that occupy the same percentile rank within the score distributions. An explicit analytical formula for PRM was provided by Braun and Holland (1982).

von Davier et al. (2004) suggested that the PRM can be viewed as a way of continuizing the discrete distributions in a similar way as (5.18). They argued that any method of equipercentile equating, including PRM and KE, must reflect the equating design and include attention to presmoothing and how to perform the estimation of observed score equating. The way that PRM and KE differ is how they continuize the discrete distributions.

Holland and Thayer (1989) demonstrated that PRM is fundamentally grounded in the convolution of the score probability distribution of the random variable X and the distribution function of a uniform random variable. In fact, consider a continuized random variable $X^* = X + V$, where $V \sim U(-0.5, 0.5)$ and U represents the uniform distribution. Then,

$$F_{X^*}(x) = \Pr(X^* \leq x) = \sum_j F_V(x - x_j) r_j \tag{5.49}$$

$$= \sum_{i \in A} r_i + r_j(x - x_j + 0.5) , \quad x_j - 0.5 \leq x \leq x_j + 0.5, \tag{5.50}$$

where $A = \{i : x_i \leq x - 0.5\}$. Evaluating $F_{X^*}(x)$ at x_j, we obtain

$$F_{X^*}(x_j) = \sum_{i < j} r_i + \frac{1}{2} r_j , \tag{5.51}$$

which can be read as the probability of scoring below a score x_j plus half the probability of scoring at score x_j. This is exactly the definition for the percentile rank of a score x_j, which states that the percentile rank of x_j is the percentage of test takers who earned scores below x_j plus half the percentage of test takers who earned the score x_j. This fact explains why the method is called PRM. A more elaborated discussion on the PRM and how it differs from KE can be found in von Davier et al. (2004, Section 6.2).

5.5 Cumulants

In the preceding sections, we have observed that the continuized score random variables $X(h_X)$ and $Y(h_Y)$ preserved certain distributional characteristics from X and Y. Specifically, we have demonstrated that the first two moments (mean and variance) of the score random variables (or their distributions) are preserved in the continuized variables. However, assessing how well $F_{h_X}(x)$ approximates $F_X(x)$ solely based on the first two moments might be insufficient. A more thorough evaluation involves assessing how different the higher moments of $X(h_X)$ and X are.

To address this, we turn to the theory of cumulants of a probability distribution (Fisher and Wishart, 1932) which is discussed extensively in Kendall and Stuart (1997). Cumulants can be used to compare distributions, similar to how moments are used to compare distributions. Note that, if two probability distributions possess identical cumulants, they will also have identical moments, and vice versa. Notably, the first cumulant corresponds to the mean, the second cumulant to the variance, and the third cumulant is equivalent to the third central moment. Higher cumulants, however, do not have simple relationships with higher moments.

One compelling reason to employ cumulants instead of moments is that they sometimes simplify a problem. For instance, the cumulants for any normal distribution are zero for all cumulants of order three or higher. Also, the cumulants of the sum of two independent random variables are the sum of the cumulants of the added variables.

The use of cumulants within KE was first introduced by Holland and Thayer (1989) and then discussed in von Davier et al. (2004). Here, we will illustrate, through an exploration of cumulant theory, the relationship between the cumulants of the discrete variable X, those of the continuous random variable V, and those of the resulting continuous variable $X(h_X)$. We start by defining the moment generating function (MGF, Casella and Berger, 2002) of a random variable X as

$$M_X(t) = \mathrm{E}(\exp\{tX\}), \tag{5.52}$$

if the expectation exists for t in some neighborhood of 0.

Next, the cumulant generating function is defined as the natural logarithm of the MGF and can be written in a Maclaurin series expansion (Beyer, 1982) as

$$\mathcal{K}(t) = \log(M(t)) = \sum_{\eta=1}^{\infty} \frac{\kappa_\eta t^\eta}{\eta!}, \tag{5.53}$$

where κ_η is the ηth cumulant. Using the definition in (5.52), the MGF of the continuized random variable $X(h_X)$ is calculated as

$$M_{X(h_X)}(t) = E[\exp(tX(h_X))] \tag{5.54}$$
$$= E[\exp(t\{a_X(X + h_V V) + (1 - a_X)\mu_X\})] \tag{5.55}$$
$$= \exp\{t(1 - a_X)\mu_X\}E[\exp(ta_X X + ta_X h_X V)] \tag{5.56}$$
$$= \exp\{t(1 - a_X)\mu_X\}E[\exp(ta_X X)]E[\exp(ta_X h_X V)] \tag{5.57}$$
$$= \exp\{t(1 - a_X)\mu_X\}M_X(ta_X)M_V(ta_X h_X) . \tag{5.58}$$

Using the definition (5.53), the cumulant generating function of $X(h_X)$ is

$$\log(M_{X(h_X)}(t)) = t(1 - a_X)\mu_X + \log(M_X(ta_X)) + \log(M_V(ta_X h_X)) \tag{5.59}$$
$$= t(1 - a_X)\mu_X + \mathcal{K}_X(ta_X) + \mathcal{K}_V(ta_X h_X). \tag{5.60}$$

If $\kappa_{\eta,X}$, $\kappa_{\eta,V}$ and $\kappa_\eta(h_X)$ denote the cumulants of X, V, and $X(h_X)$ respectively, then

$$\mathcal{K}_{X(h_X)}(t) = (\mu_X - a_X\mu_X + \kappa_{1,X}a_X)t + (\kappa_{2,X} + \kappa_{2,V}h_X^2)a_X^2\frac{t^2}{2} + \tag{5.61}$$
$$\sum_{\eta \geq 3} a_X^\eta(\kappa_{\eta,X} + \kappa_{\eta,V}h_X^\eta)\frac{t^\eta}{\eta!}.$$

From here, and observing in Eq. (5.53) that the cumulants are the coefficients of $\frac{t^\eta}{\eta!}$, it can be seen that for $\eta \geq 3$,

$$\kappa_\eta(h_X) = a_X^\eta(\kappa_{\eta,X} + \kappa_{\eta,V}h_X^\eta). \tag{5.62}$$

This result appeared as Theorem 10.6 in Lee and von Davier (2011) without any proof.

In the particular case when $V \sim N(\mu_V, \sigma_V^2)$, it is well known that

$$M_V(t) = \exp\{\mu_V t + \frac{1}{2}\sigma_V^2 t^2\}, \tag{5.63}$$

so that the cumulant distribution function becomes $\mathcal{K}_V(t) = \mu_V t + \frac{1}{2}\sigma_V^2 t^2$. Differentiation and evaluation at $t = 0$ show immediately that $\kappa_{1,V} = \mu_V$, $\kappa_{2,V} = \sigma_V^2$ and that for $\eta \geq 3$, $\kappa_{\eta,V} = 0$. Replacing these values in Eq. (5.61) and using the fact that $\kappa_{1,X} = \mu_X$, $\kappa_{2,X} = \sigma_X^2$, for the case when V is a standard normal distribution, Eq. (5.62) becomes

$$\kappa_\eta(h_X) = a_X^\eta \kappa_{\eta,X}, \tag{5.64}$$

which appears as Theorem 4.4 in von Davier et al. (2004).

For an example when cumulants have been used within KE as an evaluation tool, and for the derivation of cumulants for logistic and uniform distributions, refer to Lee and von Davier (2011).

5.6 Summary

In this chapter, we have demonstrated how it is possible to perform *continuization*; i.e., to generate continuous approximations of discrete score distributions, through convolutions. We have also established that kernel estimators can be derived as outcomes of convolution operations.

The general derivation for the continuized score random variable provides us the flexibility to experiment with various continuous distributions to be used as kernels. In this chapter, we introduced five distinct kernels. To evaluate the quality of these approximations, one possibility is to use cumulants. The chapter ended with a derivation of a general formula that establishes the relationship between the cumulants of a discrete score random variable, those of the continuous random variable, and those of the resulting continuous variable.

6

Bandwidth Selection

In Chapter 5, we introduced continuization as the primary strategy for obtaining a continuous approximation of the score distributions. We demonstrated that the cumulative distribution function (CDF) of the continuized score random variable corresponds to a kernel estimator. The continuized score variable was defined as a weighted sum of the score random variable and a continuous random variable. The weight assigned to the continuous random variable, referred to as the bandwidth, serves as a smoothing parameter that regulates the level of smoothness in the continuization process.

Given that we are working with continuous approximations of discrete score distributions, it is crucial to select the bandwidth parameter judiciously. The choice of bandwidth should aim to ensure that the estimated distribution preserves as many characteristics as possible from the original discrete test score distributions but still provides smoothness. An advantage of using kernels with bandwidths in the continuization is that it gives us flexibility. We can, for example, select an optimal bandwidth that addresses differences between equating transformations, or select a large bandwidth that yields a linear equating, or select a small bandwidth which yields an equipercentile equating.

In this chapter, we delve into various bandwidth selection methods within the context of KE. Several bandwidth selection methods proven effective within KE are considering the probability density function (PDF), including the penalty method (von Davier et al., 2004), the rule-based method (Andersson and von Davier, 2014), double smoothing (Häggström and Wiberg, 2014), and a likelihood method (Liang and von Davier, 2014). Additionally, we explore a recently proposed variation of the penalty method and the utilization of cross-validation in KE (Wallin et al., 2018, 2021).

Almost all methods described in this chapter are based on discrepancy measures involving score probabilities and PDF or continuous approximation of the score probability mass function (PMF). Although the densities associated with score distributions are important and we therefore try to preserve characteristics of the score distributions, they are not directly used when computing the equating transformation. Instead, the CDF has an important role. Thus, the chapter ends with a suggestion for an innovative approach that seeks to develop a bandwidth selection method based on the CDF rather than the PDF of scores.

6.1   Bandwidth Selection in Kernel Density Estimation

For the general definition of a kernel density estimate provided in Equation (5.1), consider for a moment that the graphical representation of the function $k(\cdot)$ resembles a hill. Using this analogy, we can view the density estimator as a weighted sum of these hills, with each hill centered at an observation. The kernel function defines the shape of these hills, while the bandwidth determines their width.

In order to evaluate the effect that the bandwidth h_X has on the estimator, let us examine extreme values of it. If h_X is very small (tending to 0), the hills will be extremely narrow, and the resulting density estimate will resemble a sum of spikes at each observation. This leads to a density that is not very smooth. Conversely, if h_X is large (approaching infinity), the hills become flattened, obscuring the finer details of their shape, and the estimate becomes smoother.

A visual representation of this type of analysis is provided in Figure 6.1, where the density estimate for a hypothetical sample is displayed for different values of h_X, taking the values 0.1, 0.3, 0.7, and 1.0. The key takeaway from this analysis is that the selection of the bandwidth parameter is of crucial importance for an appropriate estimation of the density.

While numerous bandwidth selection methods are available in the kernel density estimation literature, the focus in this chapter is on discussing methods that are well-suited for integration within the GKE framework. Irrespective of the specific method chosen, it is essential to keep in mind that the bandwidth selection process should prioritize the preservation of the original discrete score distributions to the greatest extent possible.

6.2   Selecting Bandwidths

Most of the methods that have been proposed to select the bandwidth are based on a measure of the discrepancy between the unknown density and the estimator. Some common choices of discrepancy measures are the mean squared error (MSE), the integrated squared error (ISE), and the mean integrated squared error (MISE). The MSE is calculated pointwise, i.e., for each data point, and when including the bandwidth h_X the MSE can be defined as

$$\text{MSE}(h_X) = \text{E}\{\hat{f}_{h_X}(x) - f(x)\}^2. \tag{6.1}$$

A measure of the global performance of the estimator is the ISE, defined as

$$\text{ISE}(h_X) = \int \{\hat{f}_{h_X}(x) - f(x)\}^2 dx. \tag{6.2}$$

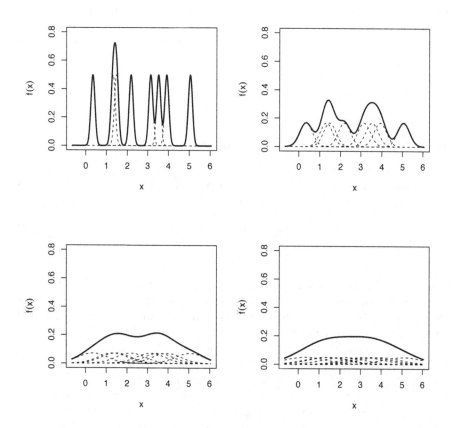

FIGURE 6.1: Kernel density estimates for a hypothetical sample. From top left to right, and from bottom left to right, $h_X = 0.1, 0.3, 0.7, 1.0$.

A modification of MSE that leads to a global measure of accuracy is the MISE (Rosenblatt, 1956), defined as

$$\text{MISE}(h_X) = \text{E} \int \{\hat{f}_{h_X}(x) - f(x)\}^2 dx . \tag{6.3}$$

The value of h_X that minimizes these measures is selected as the optimal bandwidth. Methods such as the rule-of-thumb and least squares cross-validation are derived from these measures of discrepancy. The interested reader is referred to Silverman (1986) for more details on these methods. Regardless of the chosen discrepancy measure, it has been proven that the selected bandwidths lead asymptotically to the same level of smoothing (Härdle, 1991).

Although appealing, it is worth noting that not all bandwidth selection methods are appropriate to be used in the context of KE. In fact, all these methods select the bandwidth by comparing the estimated density with the real density. In the context of KE, however, only the continuized score variable has an associated density, but no real density exists for the discrete scores. This problem can be handled using penalty functions, which are described in the following section.

6.3   Minimizing a Penalty Function

In general optimization problems, penalty functions are used to incorporate constraints into the objective function (e.g., Smith and Coit, 1997). For instance, consider the following optimization problem

$$\min_{w} f(w) = \exp(-w) \tag{6.4}$$

$$\text{subject to} \quad w \leq c,$$

where $c > 0$. In this problem, it is easy to conclude that $w = c$ is the minimizer. Now, convert the restriction in the form $w - c \leq 0$ and set *penalties* each time the constraint is violated. Since we are minimizing a function, we could add a positive value each time that $w - c > 0$ and add nothing if the constraint is satisfied. A penalty function that can be used is, for instance, $\text{PEN}(w) = \max(0, w - c)$. If the restriction is of the form $g(w) = c$, for some function g, we can use $\text{PEN}(w) = (g(w) - c)^2$. In this case, note that the lowest value of the penalty is achieved when $g(w) = c$ and PEN is equal to 0. This is exactly what we need, i.e., adding 0 when the restriction is satisfied. Thus, the restriction translates to the optimization problem of minimizing the penalty function.

Minimizing a penalty function was the first approach used in KE for automatic selection of the bandwidth (von Davier et al., 2004). The main idea is that the (estimated) continuous version of the score probability distribution evaluated at each score point should be very similar and, in the ideal case, equal to, the (estimated) discrete score probabilities. Thus, under the restriction that $\hat{r}_j = \hat{f}_{h_X}(x_j)$, the penalty method selects h_X as the value that minimizes

$$\text{PEN}_1(h_X) = \sum_{j} \left(\hat{r}_j - \hat{f}_{h_X}(x_j) \right)^2. \tag{6.5}$$

Further characteristics of the smooth densities can be accounted for by considering a second penalty,

$$\text{PEN}_2(h_X) = \sum_{j} A_j, \tag{6.6}$$

where A_j is an indicator variable taking the value 1 if $\hat{f}'_{h_X}(x_j - v) < 0$ and $\hat{f}'_{h_X}(x_j + v) > 0$, and 0 otherwise. Here, $\hat{f}'_{h_X}(x_j)$ is the derivative of $\hat{f}_{h_X}(x_j)$ and the parameter v indicates the width of the interval around the score x, often set to 0.25 (Lee and von Davier, 2011; von Davier, 2013).

Combining the two penalty functions by weighting PEN$_2$ by the constant κ, we obtain

$$\text{PEN}(h_X) = \sum_j \left(\hat{r}_j - \hat{f}_{h_X}(x_j)\right)^2 + \kappa \cdot \sum_j A_j. \qquad (6.7)$$

The first term in Eq. (6.7) (PEN$_1$) is used to ensure that the characteristics of the score distribution are preserved by the continuized density function at each score value. Because the first term might undersmooth the data, the smoothness of the continuized distribution is ensured by using the second term (PEN$_2$).

6.4 Leave-One-Out Cross-Validation

Cross-validation (CV) is a general method used for the evaluation of models. The main idea in CV is to use only part of the data, for instance, half the data or various subsets of the data, to estimate a model so that the removed data can be used to evaluate the fitted model. The extreme version of the method considers each observation as a subset and is known as leave-one-out cross-validation (LCV) (Stone, 1974). In the context of kernel density estimation, LCV is used when the density estimate is computed using all the data except the ith.

The LCV method was proposed to be used within the KE framework by Wallin et al. (2018, 2021). In LCV applied to KE, the bandwidth is selected as the value that minimizes

$$\text{LCV}(h_X) = \frac{1}{J} \sum_{j=1}^{J} (\hat{r}_j - \hat{f}_{h_X}^{(-j)}(x_j))^2, \qquad (6.8)$$

where

$$\hat{f}_{h_X}^{(-j)}(x) = \sum_{l=1, l \neq j}^{J} \hat{r}_l K\left(\frac{x - \hat{a}_X x_l - (1 - \hat{a}_X)\hat{\mu}_X}{\hat{a}_X h_X}\right) \frac{1}{\hat{a}_X h_X}, \qquad (6.9)$$

is an estimate based on the subsample when $(x_j, \hat{r}_j)$ is left out.

A straightforward extension of LCV that is based on the penalty method in Eq. (6.7) was proposed by Wallin et al. (2021). These authors extended the LCV method by adding PEN$_2(h_X)$ to the objective function, leading to what they called the penalty leave-one-out CV (PLCV). Using the PLCV method, the bandwidth is chosen as the value that minimizes

$$\text{PLCV}(h_X) = \frac{1}{J} \sum_{j=1}^{J} (\hat{r}_j - \hat{f}_{h_X}^{(-j)}(x_j))^2 + \kappa \cdot \sum_{j} A_j. \tag{6.10}$$

6.5 Likelihood Cross-Validation

Likelihood cross-validation is another general method used to evaluate the fit of a statistical model. In the context of kernel density estimation, the idea is to view the density estimate as a function of the bandwidth h_X for fixed values of the data, so that for a new data point that is independent of the original sample, the density estimate can be considered as the likelihood of h_X. In practice, the new data point is replaced by one from the observed sample. For a random sample $X_1, \ldots, X_N$, the maximum likelihood cross-validation (MLCV) method selects the bandwidth by maximizing

$$\text{MLCV}(h_X) = \prod_{i=1}^{N} \hat{f}_{h_X}^{(-i)}(x_i), \tag{6.11}$$

where $\hat{f}_{h_X}^{(-i)}(x)$ has been obtained with observation x_i omitted. A particular case of MLCV was proposed by Liang and von Davier (2014) for use within KE. The authors refer to this method as CV, but to avoid confusion with traditional CV and the LCV method described in Section 6.4, we use the term likelihood cross-validation (LiCV) when referring to this method.

The first step in this method is to obtain two subsamples by randomly splitting the data. Let N_{X_j} be the random variable that represents the number of test takers scoring $X = x_j$, $j = 1, \ldots, J$, with corresponding realized value denoted as n_{X_j}. In the LiCV, the frequencies are assumed to be count data such that $N_{X_j} \sim \text{Poisson}(\hat{f}_{h_X}(x_j))$, where the values $\hat{f}_{h_X}(x_j)$ are calculated using the first random subsample of the data, and the observed n_{X_j} come from the second subsample. The LiCV method initially selects the value of h_X that maximizes

$$\text{LiCV}(h_X) = \prod_{j=1}^{J} \frac{\exp\{-\hat{f}_{h_X}(x_j)\}(\hat{f}_{h_X}(x_j))^{n_{X_j}}}{n_{X_j}!}. \tag{6.12}$$

The random split of the data set and the bandwidth selection are repeated many times, and the median of the obtained bandwidths is selected as the optimal bandwidth. An example when this method is used and compared with other bandwidth selection methods in KE can be found in Wallin et al. (2021).

6.6 Double Smoothing

Double smoothing (DS, Hall et al., 1992) is a general procedure used to select the bandwidth parameter that is based on the minimization of an estimate of the targeted estimator's MSE. It is typically divided into two steps. In the first step, a pilot bandwidth is used to estimate the part of the MSE that contains the bias. Next, the method selects the bandwidth that minimizes the MSE, where the bias component has been approximated in the previous step. Thus, the first step includes a *single* smoothing of the bias, whereas in the second the *double* smoothing is completed. Further details about DS in general are provided in Hall et al. (1992).

A version of DS that has been shown to be a possible choice when selecting the bandwidths in KE, was proposed by Häggström and Wiberg (2014). In the first step, a pilot bandwidth, g_X, is used to estimate the score density such that

$$\hat{f}_{g_X}(x) = \sum_{j=1}^{J} \hat{r}_j \phi \left(\frac{x - \hat{a}_X^{g_x} x_j - (1 - \hat{a}_X^{g_X}) \hat{\mu}_{XT}}{g_X \hat{a}_X^{g_X}} \right) \frac{1}{g_X \hat{a}_X^{g_X}}, \tag{6.13}$$

with

$$\hat{a}_X^{g_X} = \sqrt{\hat{\sigma}_{XT}^2 / (\hat{\sigma}_{XT}^2 + g_X^2)}. \tag{6.14}$$

In the second step, $f_{h_X}(x)$ is estimated using an augmented score sample, x^*, which contains the observed score sample $x = (x_1, \ldots, x_J)$, and the values halfway between them, so that $x^* = (x_1, x_1 + 0.5, x_2, \ldots, x_J - 0.5, x_J)^t$. In this estimation, the r_j values are replaced by $\hat{f}_{g_X}(x_j)$, leading to

$$\hat{f}_{h_X}^*(x) = \sum_{j=1}^{J} \hat{f}_{g_X}(x_j) \phi \left(\frac{x - \hat{a}_X x_j - (1 - \hat{a}_X) \hat{\mu}_{XT}}{h_X \hat{a}_X^{h_X}} \right) \frac{1}{h_X \hat{a}_X^{h_X}}, \tag{6.15}$$

where $\phi(z)$ is the standard normal density function.

The aim of the final DS estimate is to preserve the characteristics of the estimated score frequency distributions, and it can thus be considered to be another method based on a penalty function (see Section 6.3). As a matter of fact, the resulting objective function is similar to PEN$_1$, which is used in the penalty method described in Eq. (6.7), and reads as

$$\text{DS}(h_X) = \sum_{l=1}^{2J-1} (\hat{r}_l^* - \hat{f}_{h_X}^*(x_l^*))^2. \tag{6.16}$$

where $l = 1, \ldots, 2J - 1$ and

$$\hat{r}_l^* = \begin{cases} \hat{r}_{\frac{l+1}{2}} & \text{if } l \text{ is odd,} \\ \hat{f}_{h_X}^*(x_l^*), & \text{if } l \text{ is even,} \end{cases} \tag{6.17}$$

To perform DS in KE, the following steps are followed:

1. Begin with a very smooth estimate of the density function. This is achieved by using a large subjectively chosen bandwidth g_X and estimating f_{g_X} at $\boldsymbol{x}^*$; i.e., the score values and at the values halfway between the score values, $\boldsymbol{x}^* = \{x_l^*\} = (x_1, x_1 + 0.5, x_2, \ldots, x_J - 0.5, x_J)^t$, $l = 1, \ldots, 2J - 1$. (see Eq. (6.13)).

2. The obtained smooth estimate $\hat{f}_{g_X}$ will not perfectly interpolate the estimated score probabilities, the $\hat{r}$:s. We can improve the first estimate by estimating f_{h_X} at $\mathbf{x}^*$ using $\hat{f}_{g_X}$ at the actual score values, $\mathbf{x}$, and obtain a DS estimate $\hat{f}_{h_X}^*$ (see Eq. (6.15)).

3. Finally, we select the bandwidth h_X that minimizes the sum of the squared difference between $\hat{r}_l^*$ and the lth DS estimate, $\hat{f}_{h_X}^*(x_l^*)$ (see Eq. (6.16)).

At this point, we want the reader to notice that when comparing bandwidth selection methods based on a penalty function in KE, no differences were found with respect to equated scores, standard errors, and mean squared errors, when including or excluding the use of PEN_2 in the penalty function (Wallin et al., 2021). Further, Häggström and Wiberg (2014) only observed small differences in the equated scores at the end points under the EG design but no differences under the NEAT design when including PEN_2 in the penalty method. PEN_2 only seems to matter in cases with irregular shapes or if presmoothing has not been used (von Davier et al., 2004).

6.7   Rule-Based Bandwidth Selection

The MISE (see Eq. (6.3)) was defined as an evaluation measure for a density estimator. In kernel density estimation, a general method for selecting bandwidth is to minimize the asymptotic mean integrated squared error (AMISE) (Silverman, 1986; Jonas et al., 1996)

$$\text{AMISE}(h_X) = (Nh_X)^{-1} R(k) + h_X^4 R(f'') \left[\int x^2 k/2 dx \right]^2, \tag{6.18}$$

where f denotes the true density function to be estimated, k is the used kernel and $R(\theta) = \int \theta(x)^2 dx$. The AMISE is minimized with respect to the bandwidth h_X by

$$h_X = \left[\frac{R(k)}{NR(f'')(\int x^2 k)^2}\right]^{1/5} \tag{6.19}$$

(Scott, 1992). When using a standard normal kernel for an underlying normally distributed random variable X, an approximation, known as Silverman's rule of thumb (SRT) is obtained as

$$h_X \approx 1.06\sigma_X N_X^{-1/5}, \tag{6.20}$$

where N_X is the sample size and σ_X is the standard deviation. For more details, refer to Scott (1992).

Although test scores typically are discrete, Andersson and von Davier (2014) argued that they often take the shape of normally distributed data, but with a restricted range that reduces the effect of outliers that could influence the data improperly. Their solution was to propose an adjusted SRT with a standard Gaussian kernel. Their modification of SRT leads to the selection of the optimal bandwidth parameter in KE defined as

$$\text{SRT}(h_X) = \frac{9\sigma_X}{\sqrt{100N_X^{2/5} - 81}}. \tag{6.21}$$

For details of how the SRT was derived and how it works in different situations, refer to Andersson and von Davier (2014).

6.8   A CDF-Based Bandwidth Selection Method

Up to this point, the methods we have described are based on discrepancy measures involving score probabilities and the density of the continuous approximation of the score distribution. Although the densities associated with score distributions are important and we have tried to preserve characteristics of the score distributions on them, when it comes to computing the equating transformation, the CDFs play a central role. Therefore, it appears natural to explore bandwidth selection methods that directly utilize CDFs for this purpose.

In the field of nonparametric estimation of CDFs, various methods have been proposed (see, e.g., del Roo and Estevez-Perez, 2012, and references therein). Nevertheless, since these methods are extensions of those used in kernel density estimation, they are based on the measures described in Section

6.2 and are not directly adaptable for use in equating. Based on the method described in Section 6.3, we suggest using a penalty function of the form

$$\text{PEN}(h_X) = \sum_j \left(\hat{F}(x) - \hat{F}_{h_X}(x_j) \right)^2, \tag{6.22}$$

where $\hat{F}(x) = \sum_{j, x_j \leq x} \hat{r}_j$. We expect this bandwidth selection method to be useful in GKE as it is built on the CDFs. The further exploration of this bandwidth selection suggestion, along with comparisons to other bandwidth methods described in this chapter is, however, left for future research.

6.9 Summary

In this chapter, we have covered various bandwidth selection methods that have been employed within the context of KE, including some recent approaches. Additionally, we gave a suggestion of a novel bandwidth selection method that departs from previous KE methodologies by being based on the CDF rather than the PDF. Methods grounded in the CDF are expected to be theoretically appealing because the equating transformation is built upon the CDF. However, this has yet to be examined and research is needed to explore the potential of this suggestion.

While most of the bandwidth methods discussed here have been examined and compared in a study by Wallin et al. (2021), a comprehensive comparison involving the new proposed CDF method remains unexplored and is left as a subject for future research.

7

Equating

Test equating is a statistical process that is used to adjust for test form differences in standardized testing programs in which the test forms have been assembled according to the same specifications and have been administered under similar (standardized) conditions. The primary objective of test equating is to make it a matter of indifference for a test taker which test form he or she has taken. This allows the comparison of scores for individuals who have taken different test forms or for groups of individuals who have taken different test forms on different occasions or under different conditions. The equating transformation function plays a crucial role in this process, serving to adjust the score scale from one test form to enable a direct comparison with a score scale from another test form. A general definition of the test equating function was presented in Chapter 1.

In this chapter, we start by discussing assumptions needed for equating. We will next discuss score comparability between test forms briefly, before providing the explicit mathematical expressions of the equating transformation within the context of GKE. When describing the equating transformation, we want to highlight that it is dependent on the chosen presmoothed model (Chapter 3), which yields different estimated score probabilities (Chapter 4). These affect the equating transformation together with which continuization (Chapter 5) has been used, including the chosen bandwidth method (Chapter 6). The GKE approach has been demonstrated to offer flexibility to accommodate various forms of equating transformations, including linear versions and chained versions that result from composing two or more equating transformations.

We present the mathematical expressions for equating transformations within three KE frameworks: the observed-score equating (OSE) framework, the IRT framework, and the local equating framework. To define the latter, we introduce the equating transformation utilized in local equating, thereby introducing the concept of the local KE framework. The chapter ends with a brief discussion of how to graphically represent the equating transformation.

7.1   Assumptions

The chapter devoted to equating in Lord (1980) essentially conveyed the notion that equating is either infeasible or unnecessary. This perspective implicitly acknowledges the alignment between two extremes: the ideal scenario for equating, where all assumptions are met, resulting in two testing instruments providing identical score distributions, and the practical challenges encountered in real-world applications, where several assumptions may not hold. In Section 1.1, the fundamental requirements of equating were listed. To gain a more detailed and explicit understanding, we look more closely here at the assumptions that must be satisfied for a meaningful test equating procedure.

A. The fundamental assumption inherent to all equating methods is the equal construct requirement, which means that the two tests being compared are considered "equivalent." In this context, equivalence implies that both tests are measuring the same underlying construct or ability. This assumption holds significant importance because it provides the confidence to compare scores from two different test forms, knowing that any score differences can be attributed to variations in test difficulty rather than discrepancies in content or construct being measured.

The equal construct requirement aligns seamlessly with the framework of IRT equating, in which we assume that the same underlying construct is being assessed by both testing instruments, and thus the same IRT model is typically applicable to both sets of data. A consequence of this assumption is the necessity for standardization in the administration of tests across different testing instruments, as alterations in the test-taking experience can potentially influence how test takers respond to items.

B. The equal reliability requirement states that two test forms being equated should have equal reliability. In practice, this usually means that the test forms that are equated should not differ too much in length, as that typically means different reliability for tests scored by number correct scoring rules or an 1PL IRT model. This also means that if some items must be excluded due to being mistakenly leaked to the public, it might be problematic to equate that test form with a test form that can keep all items because the items have not been leaked.

C. The symmetry requirement states that the transformation mapping scores from an origin test scale to their equivalents on a target test scale should be the inverse of the transformation used to map the scores in the reverse direction, effectively reversing the order of the scales. In practical terms, this implies that various regression functions are eliminated as potential equating transformation functions. To assess the validity of this requirement, one approach involves equating test scores from one scale to another and subsequently equating the scores back to the original scale. Alternatively, one may conduct equating in a circular fashion, beginning by equating test scores from the original scale to scale A, then equating scores at scale A to another scale B, and finally equating scores from scale B back to the original scale. There are problems, however, with equating in a circle and for a discussion see, for example, Brennan and Kolen (1987). The idea with both approaches is that the equating transformation should yield the original test scale if implemented accurately and if the symmetry requirement holds.

D. The equity requirement states that it should be a matter of indifference to a test taker which test form he or she answered. A way to achieve equity is to use local equating transformations, as will be described in Section 7.7. A fundamental assumption for this requirement to hold is that the test score equating function is monotonically increasing, as this ensures that the test scores in the scale are ordered. If complete equity is achieved, the test takers are ordered identically in both the original score scale and the equated score scale.

E. The population invariance requirement states that all test score equating functions should exhibit an approximate invariance across subgroups within the target population. Population invariance should be checked particularly among major demographic subgroups. Strategies for evaluating the population invariance of OSE methods are outlined in Dorans and Holland (2000).

F. There must either be a sufficient number of common test takers taking both tests to be equated, or there must be a sufficient number of common items included alongside the tests to be equated, or there should be sufficient information available in the form of covariates about the test takers. This assumption lies at the heart of how data collection is planned and how equating designs, as discussed in Chapter 1, are formulated to ensure the best possible opportunity for an unbiased comparison of testing instruments.

We also assume that the samples of test takers used in equating are representative of the larger test-taking population. Additionally, we assume that the common items employed in equating are representative of the various item types found on the tests, and that the information collected about the test takers is both informative and sufficient. Adhering to this assumption is essential for guaranteeing that the equating results hold validity for subgroups within the target population.

G. Among the implicit assumptions related to the target population in equating, we assume that both tests to be equated share the same target population of potential test takers. This assumption has several implications for data collection designs:

i) In the EG design, we actually assume that the two groups being equated are randomly drawn from the same populations. This assumption can and should be tested, as demonstrated in, for example, Lyrén and Hambleton (2011).

ii) In the NEAT design, we make assumptions about the conditional score distributions given the anchor test scores in the two non-equivalent samples. These assumptions, however, are untestable, which can pose problems in terms of the identifiability and accuracy of equating results. The equating transformation utilizes an artificial testing population with assigned weights for the two samples. In practice, these weights do not seem to significantly impact the equating results, except perhaps in cases of extreme imbalance (San Martín and González, 2022).

iii) To simplify the presentation, many research papers and books typically describe equating between only two test forms from two similar administrations. However, this is a simplification, as most educational testing programs involve multiple test administrations within a given year, and test results may exhibit seasonal effects. While there have been research studies on how to account for seasonality in testing samples (as seen in Dorans and Holland, 2000; Qian et al., 2013; Li et al., 2012), psychometricians must handle these assumptions and adjustments in ways that are most suitable for the specific test being equated.

H. All equating methods rely on the assumptions of the analysis frameworks they employ. These assumptions are foundational to the respective analysis frameworks and are essential for the accurate application of equating methods within those frameworks.

In the classical test theory framework, the following assumptions are made:

i) Test scores are subject to measurement error.

ii) The measurement error is random.

iii) Measurement error is independent of the true score being measured.

iv) The test score is a linear combination of the true score and an error term.

In the IRT framework, fundamental assumptions that must be satisfied include:

i) Unidimensionality: It is assumed that the items on a test are measuring a single latent trait or ability.

ii) Local independence: It is assumed that the responses to different items are independent of each other, given the latent trait.

iii) Parameter invariance: It is assumed that the (items and persons) parameters are similar regardless of which group they are estimated in.

iv) Monotonicity: It is assumed that the probability of a correct response to an item is an increasing monotonic function of the latent trait.

7.2   Score Comparability through Scale Transformation

Definition 1 establishes that any function satisfying the symmetry requirement that is capable of mapping the score scale of one test form to the scale of another test form can be used with the aim to treat the scores of these different test forms as comparable. The main consequence of transforming either of the score scales into a "converted" score scale is that the distributions functions of the score random variables defined on the unconverted and converted scales will become identical (see Lord, 1950; van der Linden, 2022). An explicit mathematical expression for the equating transformation can thus be obtained by setting these two score distributions functions equal to each other. In the next section, we will see that using this strategy, the resulting equating transformation function will also satisfy the definition given by Angoff (1971) (see also Lord, 1950), which states that "two scores, one on Form X and the other on Form Y [...] may be considered equivalent if their corresponding percentile ranks in any given group are equal."

7.3   Equating Transformations for the OSE Framework

Let $X \in \mathcal{X}$ and $Y \in \mathcal{Y}$ be the random variables representing the scores of test forms X and Y, respectively; and let F_X and F_Y be their corresponding distribution functions. Suppose there exists a function φ such that $F_Y(t) = F_{\varphi(X)}(t), \forall t \in \mathbb{R}$. The distribution function of the random variable $\varphi(X)$ can be obtained as

$$
\begin{aligned}
F_{\varphi(X)}(t) &= \Pr(\varphi_Y(X) \leq t) \\
&= \Pr(X \leq \varphi^{-1}(t)) \\
&= F_X(\varphi^{-1}(t)).
\end{aligned}
$$

Then, the condition $F_Y(t) = F_{\varphi(X)}(t)$ can equivalently be written as $F_Y(t) = F_X(\varphi^{-1}(t))$. Replacing $t = \varphi(X)$ it follows that

$$
\begin{aligned}
F_Y(\varphi(X)) &= F_X(\varphi^{-1}(\varphi(X))) \\
F_Y(\varphi(X)) &= F_X(X) \\
\varphi(X) &= F_Y^{-1}(F_X(X)),
\end{aligned}
$$

which is formalized as Theorem 1 in Braun and Holland (1982).

On the other hand, Angoff's definition can mathematically be established as follows: the X scores and the Y scores are comparable if a unique $y \in \mathcal{Y}$ exists for each $x \in \mathcal{X}$ such that

$$
F_X(x) = p = F_Y(y), \ p \in [0, 1].
$$

Assuming that both F_X and F_Y are strictly increasing functions, and by eliminating p, the equating transformation $\varphi : \mathcal{X} \mapsto \mathcal{Y}$ becomes

$$
\mathcal{Y} \ni y = \varphi(x) = F_Y^{-1}(F_X(x)) \quad \forall x \in \mathcal{X}.
$$

This expression was referred to as the equipercentile transformation in Chapter 1. The equating transformation reversing the scales mapping can be defined similarly. In fact, $\varphi^* : \mathcal{Y} \mapsto \mathcal{X}$ is defined as

$$\varphi^*(y) = F_X^{-1}(F_Y(y))$$
$$= (F_X^{-1} \circ F_Y)(y),$$

where the symbol $\circ$ denotes the composition of functions. Note that $\varphi^* = F_X^{-1} \circ F_Y$ has inverse $\varphi^{*-1} = F_Y^{-1} \circ F_X$ so that

$$\varphi^{*-1}(x) = (F_Y^{-1} \circ F_X)(x)$$
$$= F_Y^{-1}(F_X(x))$$
$$= \varphi(x),$$

and $\varphi(x)$ thus produces a symmetric equating.

Because both F_X and F_Y are usually discrete distribution functions, they are replaced by continuous versions obtained after the continuization step introduced in Chapter 5,

$$F_{h_X}(x) = \sum_j r_j K(R_{jX}(x)),$$

$$F_{h_Y}(y) = \sum_k s_k K(R_{kY}(y)),$$

where $R_{jX}(x) = \frac{x - a_X x_j - (1-a_X)\mu_X}{a_X h_X}$, a_X is defined in (5.19), and similar definitions can be obtained for $R_{jY}(y)$ and a_Y. With such replacement, the KE transformation function is computed as defined in (2.3), which we repeat here:

$$\varphi(x; \hat{r}, \hat{s}) = F_{h_Y}^{-1}(F_{h_X}(x; \hat{r}); \hat{s}),$$

where $\hat{r}$ and $\hat{s}$ are vectors of the estimated score probabilities (Step 2 described in Chapter 4) obtained from the (possibly presmoothed) score distributions (Step 1 described in Chapter 3). Both r and s are obtained using the DF, and their coordinates are defined as $r_j = \Pr(X = x_j)$ and $s_k = \Pr(Y = y_k)$, with x_j and y_k taking values in $\mathcal{X}$ and $\mathcal{Y}$, respectively. Finally, h_X and h_Y are bandwidth parameters (described in Chapter 6) controlling the degree of smoothness for the continuization (Step 3 described in Chapter 5).

A particularly useful feature of the KE transformation is that different shapes of the equating transformation can be obtained based on different values of the bandwidth parameter. In general, a high value of the bandwidth parameter will give rise to a linear equating transformation, as will be discussed in Section 7.4, while a low value of the bandwidth parameter will give rise to the equating transformation under Angoff's definition used in the percentile rank method (see Section 5.4.6).

Recall the definition of a continuized score random variable in Eq. (5.17), where V is a continuous random variable (the kernel function of choice, in this case). In the next section, it will be explicitly shown that a linear version of the equating transformation can be obtained for particular choices of the bandwidth parameter. The following theorem will be useful to show these statements.

Theorem 3. *The following statements hold:*

a) $\lim_{h_X \to 0} a_X = 1$

b) $\lim_{h_X \to \infty} a_X = 0$

c) $\lim_{h_X \to \infty} h_X a_X = \sigma_X / \sigma_V$

d) $\lim_{h_X \to 0} X(h_X) = X$

e) $\lim_{h_X \to \infty} X(h_X) = (\sigma_X / \sigma_V)V + \mu_X$

Theorem 3, which appears as Theorem 10.2 in Lee and von Davier (2011) (which is a generalization of Theorem 4.1 in von Davier et al. (2004)), helps us understand the conditions under which the method can be used to accurately equate scores on two different test forms. In particular, this theorem illustrates what happens when the bandwidth tends toward extremes, that is, 0 or infinity. If the bandwidth is close to zero, a continuous function that closely tracks the original discrete variable is obtained. When the bandwidth goes towards infinity, the discrete variable is dwarfed, and the resulting probability function will resemble the distribution of V with the mean of the original discrete variables, and variance σ_X^2 / σ_V^2. These properties will help us achieve a family of equating functions, including members from a linear equating to an equipercentile equating, all within the GKE framework.

7.4   Linear Equating Transformation

As pointed out in Section 1.4, when the score distributions of X and Y are of the location-scale type, the equating transformation becomes a linear function. As a matter of fact, using the results of Theorem 3, it is easy to prove that

$$\lim_{h_X \to \infty} R_{jX}(x) = \left(\frac{x - \mu_X}{\sigma_X / \sigma_V} \right),$$

and thus

$$\lim_{h_X \to \infty} F_{h_X}(x) = K_X \left(\frac{x - \mu_X}{\sigma_X/\sigma_V} \right),$$

$$\lim_{h_Y \to \infty} F_{h_Y}(y) = K_Y \left(\frac{y - \mu_Y}{\sigma_Y/\sigma_W} \right),$$

where the kernels K_X and K_Y have been indexed by X and Y to explicitly show that they can be different, and W is a continuous random variable similar to V. This result shows that the continuized score probability distribution functions are of location-scale type when the bandwidth parameter is large. Moreover, if $F_{h_Y}(y) = u$, then

$$y = \mu_Y + \frac{\sigma_Y}{\sigma_W} K_Y^{-1}(u),$$

$$F_{h_Y}^{-1}(u) = \mu_Y + \frac{\sigma_Y}{\sigma_W} K_Y^{-1}(u),$$

and replacing u with $F_{h_X}(x)$, it follows that

$$F_{h_Y}^{-1}(F_{h_X}(x)) = \mu_Y + \frac{\sigma_Y}{\sigma_W} K_Y^{-1} \left[K_X \left(\frac{x - \mu_X}{\sigma_X/\sigma_V} \right) \right]$$

$$= \mu_Y + \frac{\sigma_Y}{\sigma_W} \left\{ K_Y^{-1} \left[K_X \left(\frac{x - \mu_X}{\sigma_X/\sigma_V} \right) \right] - \left(\frac{x - \mu_X}{\sigma_X/\sigma_V} \right) \right\}$$

$$+ \frac{\sigma_Y}{\sigma_W} \left(\frac{x - \mu_X}{\sigma_X/\sigma_V} \right)$$

$$= \mu_Y + \frac{\sigma_Y}{\sigma_X} \left(\frac{\sigma_V}{\sigma_W} \right) (x - \mu_X) + \frac{\sigma_Y}{\sigma_W} \Delta(z),$$

where $z = \left(\frac{x - \mu_X}{\sigma_X/\sigma_V} \right)$ and $\Delta(t) = \{K_Y^{-1}(K_X(t)) - t\}$ is the shift function (e.g., Doksum, 1974). Thus, if $\Delta(z) = 0$, $\forall z$ and $\sigma_V = \sigma_W$, it follows that

$$F_{h_Y}^{-1}(F_{h_X}(x)) = \frac{\sigma_Y}{\sigma_X} (x - \mu_X) + \mu_Y.$$

If $\Delta(z) \neq 0$, the approximation is better when $\sigma_V = \sigma_W$ and $\Delta(z)$ is small. Note that even if $\sigma_V = \sigma_W$, it can happen that $K_Y \neq K_X$ and thus $\Delta(z) \neq 0$.

7.5 Equating Transformation According to Equating Design

In Chapter 1, it was pointed out that an important aspect in the definition of the equating transformation is the fact that it has to be valid for a certain population. As such, it needs to be defined on a common population of interest, which is not always trivial for some data collection strategies. In Chapter 4, it was shown that the DF functions for the EG, SG, and CB designs lead to two vectors of score probabilities r and s, and thus only two cumulative distribution functions are needed to obtain φ. The same was true for the NEAT design, for which additional identification restrictions were needed to obtain r and s. For all of these designs, the equating transformation takes the general form given in Eq. (2.3). However, the equating transformation for chained equating is different, as described in the next subsection.

7.5.1   Chained Equating

The equipercentile transformation can be used to equate two test forms (X and Y) through a third test form. In this case, the test scores X and Y are connected or *chained* together through the third test form. Let X, Y, and Z be three different test forms and let $\mathcal{X}$, $\mathcal{Y}$, and $\mathcal{Z}$ be sample spaces. Define φ_Z and φ_Y as

$$\varphi_Z : \mathcal{X} \mapsto \mathcal{Z},$$
$$\varphi_Y : \mathcal{Z} \mapsto \mathcal{Y},$$

such that $\varphi_Z(x) = F_Z^{-1}(F_X(x))$ and $\varphi_Y(z) = F_Y^{-1}(F_Z(z))$. Then, the equating transformation that maps $\mathcal{X}$ to $\mathcal{Y}$ through $\mathcal{Z}$, which we denote as $\varphi_{(CE)}$, is defined as

$$\varphi_{(CE)} : \mathcal{X} \mapsto \mathcal{Y}, \tag{7.1}$$

with $\varphi_{(CE)} = (\varphi_Y \circ \varphi_Z)$.

To show that $\varphi_{(CE)}$ effectively maps $\mathcal{X}$ to $\mathcal{Y}$, note that

$$\begin{aligned}
\varphi_Y \circ \varphi_Z &= (F_Y^{-1} \circ F_Z) \circ \varphi_Z \\
&= F_Y^{-1} \circ (F_Z \circ \varphi_Z) \\
&= F_Y^{-1} \circ (F_Z \circ [F_Z^{-1} \circ F_X]) \\
&= F_Y^{-1} \circ (F_Z \circ F_Z^{-1}) \circ F_X \\
&= F_Y^{-1} \circ F_X,
\end{aligned}$$

where we have used the associative law of function composition twice. Thus,

$$\begin{aligned}
(\varphi_Y \circ \varphi_Z)(x) &= (F_Y^{-1} \circ F_X)(x) \\
&= F_Y^{-1}(F_X(x)),
\end{aligned}$$

which is the equating transformation that maps $\mathcal{X}$ to $\mathcal{Y}$.

In the context of KE, the score distributions F_X, F_Y, and F_Z are replaced by their continuized versions F_{h_X}, F_{h_Y}, and F_{h_Z}, respectively. The chained equating transformation will be useful when the test form Z is replaced with an anchor test form A under the NEAT design. In such case, however, two cautions must be taken: i) to solve the identifiability problem inherent to the NEAT design, ii) to define the equating transformation as valid for a common target population. To solve the identifiability problem, instead of imposing identification restrictions on the score probability parameters, as was the case for PSE (see Section 4.1.4), restrictions on the CDFs themselves are considered. In what follows, we use two identifiability restrictions that are useful to tackle both these points.

Let X be a test form that is administered to population $\mathcal{P}$, Y a test form administered to population $\mathcal{Q}$, and A an anchor test that is administered to both $\mathcal{P}$ and $\mathcal{Q}$. Define

$$\varphi_A : \mathcal{X} \mapsto \mathcal{A},$$
$$\varphi_Y : \mathcal{A} \mapsto \mathcal{Y},$$

such that $\varphi_A(x) = F_A^{-1}(F_X(x))$, and $\varphi_Y(a) = F_Y^{-1}(F_A(a))$. Then, the $\varphi_{(CE)}$ equating transformation that maps $\mathcal{X}$ to $\mathcal{Y}$ through $\mathcal{A}$ is defined as

$$\varphi_{(CE)} : \mathcal{X} \mapsto \mathcal{Y}, \tag{7.2}$$

where $\varphi_{(CE)} = (\varphi_Y \circ \varphi_A)$.

Although the equating transformation $\varphi_A : \mathcal{X} \mapsto \mathcal{A}$ can be estimated using the data coming from population $\mathcal{P}$, and $\varphi_Y : \mathcal{A} \mapsto \mathcal{Y}$ can be estimated using the data from population $\mathcal{Q}$, the equating transformation $\varphi_{(CE)}$ needs to be defined on a common population. Moreover, neither φ_A nor φ_Y are identified as they depend on the unidentified distributions functions $F_X(x)$ and $F_Y(y)$, respectively. The following identifiability restrictions are useful to tackle both these problems.

Let, F_{XP}, F_{XQ}, F_{YP}, F_{YQ} be the score probability distribution functions of X and Y in populations $\mathcal{P}$ and $\mathcal{Q}$, respectively. Similarly, define F_{AP} and F_{AQ} as the distribution functions of the anchor scores in populations $\mathcal{P}$ and $\mathcal{Q}$. Then, under the restrictions

IR-CE1: $F_{XQ}(x) = F_{AQ}(F_{AP}^{-1}(F_{XP}(x)))$,

and

IR-CE2: $F_{YP}(y) = F_{AP}(F_{AQ}^{-1}(F_{YQ}(y)))$,

$F_X(x)$ and $F_Y(y)$ become identified and $\varphi_{(CE)}$ is properly defined in both P and Q.

As test form X is only administered to population $\mathcal{P}$, and test form Y is only given to population $\mathcal{Q}$, **IR-CE1** and **IR-CE2** ensure that the unidentified distributions F_{XQ} and F_{YP} are dependent on identified distributions and thus become identified. Given that both F_{XP} and F_{YQ} are identified, then $F_X(x)$ and $F_Y(y)$ become identified.

On the other hand, to show that $\varphi_{(CE)}$ is defined on a common population, note that **IR-CE1** is equivalent to

$$F_{AQ}^{-1} \circ F_{XQ} = F_{AP}^{-1} \circ F_{XP},$$

or

$$\varphi_{AQ}(x) = \varphi_{AP}(x),$$

and so the transformation $\mathcal{X} \mapsto \mathcal{A}$ is population invariant. Analogously, **IR-CE2** is equivalent to

$$F_{AP}^{-1} \circ F_{YP} = F_{AQ}^{-1} \circ F_{YQ} \Leftrightarrow F_{YP}^{-1} \circ F_{AP} = F_{YQ}^{-1} \circ F_{AQ},$$

or $\varphi_{YP}(a) = \varphi_{YQ}(a)$, and so the transformation $\mathcal{A} \mapsto \mathcal{Y}$ is also population invariant. This means that the chained equating transformation becomes

$$\begin{aligned}
\varphi_{(CE)}(x) &= (\varphi_Y \circ \varphi_A)(x) \\
&= \varphi_Y(\varphi_A(x)) \\
&= \varphi_Y(F_{AP}^{-1}(F_{XP}(x))) \\
&= F_{YQ}^{-1}(F_{AQ}(F_{AP}^{-1}(F_{XP}(x)))).
\end{aligned}$$

Note that instead of two CDFs, we need four CDFs to compute $\varphi_{(CE)}$. This is why when a chain of transformations is used for equating under the NEAT design, it was shown in Section 4.1.4 that the **DF** leads to

$$\mathbf{DF}(\mathbf{\Pi}) = \begin{pmatrix} r_P \\ t_P \\ t_Q \\ s_Q \end{pmatrix},$$

where r_P, t_P, t_Q, and s_Q, are used for the estimation of F_{XP}, F_{AP}, F_{AQ}, and F_{YQ}, respectively.

7.6 IRT Calibration

In Section 4.3 we described how to obtain estimated score probabilities when we have parameter estimates from IRT models. In the SG and EG designs, we can calibrate both testing instruments of all the test taker samples at the same time. It is also possible to calibrate the EG samples separately. If we use exactly the same IRT model with exactly the same identifiability constraints, then the item parameters will be placed on the same scale and no additional transformation is needed, (see e.g., Barrett and van der Linden, 2019).

In a NEAT design, however, we need to transform the IRT scales as they may differ. IRT linking of item parameters is needed to place IRT parameter estimates from different calibrations of the test forms on the same scale. These methods are briefly described here, but the interested reader is referred to Chapter 6 in Kolen and Brennan (2014) for a more extensive description, and for an example of their use in GKE, refer to Chapter 10 in this book.

Assume that parameters for an IRT model are estimated for test form X from a sample from population $\mathcal{P}$ and for test form Y from a sample from population $\mathcal{Q}$. If an IRT model fits the data, then any linear transformation of the θ scale will also fit the data, provided that the item parameters are also transformed. This means that we can convert IRT parameter estimates to the same scale with a linear equation. Assume that we are using a 3PL IRT model and define, for test forms X and Y, score scales $\mathcal{X}$ and $\mathcal{Y}$ as 3PL IRT scales that differ by a linear transformation. Then, the ability scales are related with the following linear transformation

$$\theta_{yi} = A\theta_{\mathcal{X}i} + B, \qquad (7.3)$$

where A and B are IRT linking coefficients that need to be estimated and θ_{yi} and $\theta_{\mathcal{X}i}$ are the ability values for test taker i on scales $\mathcal{X}$ and $\mathcal{Y}$, respectively. Further, the item parameters from the two scales are related as follows:

$$a_{yj} = a_{\mathcal{X}j}/A, \qquad (7.4)$$
$$b_{yj} = Ab_{\mathcal{X}j} + B, \qquad (7.5)$$
$$\gamma_{yj} = \gamma_{\mathcal{X}j}, \qquad (7.6)$$

where $a_{\mathcal{X}j}$, $b_{\mathcal{X}j}$, and $\gamma_{\mathcal{X}j}$ are the item parameters for item j on scale $\mathcal{X}$ and a_{yj}, b_{yj}, and γ_{yj} are the item parameters for item j on scale $\mathcal{Y}$.

There are two general approaches for estimating A and B. The first approach contains three methods and is based on moments, i.e., the means and standard deviations of the common item parameter estimates. These methods are the mean-mean, the mean-sigma (Marco, 1977; Loyd and Hoover, 1980; Kolen and Brennan, 2014) and the mean-geometric mean (Mislevy and Bock, 1990), which is described in Ogasawara (2000). In these three methods, the means and standard deviations are defined on the common items in the NEAT design. Let $\mu_{a_\mathcal{X}}$ and $\mu_{a_\mathcal{Y}}$ be the mean of the item discrimination parameter estimates of the common items, and let $\sigma_{a_\mathcal{X}}$ and $\sigma_{a_\mathcal{Y}}$ be the corresponding standard deviations. The linking coefficient A is then calculated for the mean-mean method as

$$A = \frac{\mu_{a_\mathcal{X}}}{\mu_{a_\mathcal{Y}}}, \qquad (7.7)$$

and similarly, the coefficient A is calculated for the mean-sigma method as

$$A = \frac{\sigma_{b_\mathcal{X}}}{\sigma_{b_\mathcal{Y}}}. \qquad (7.8)$$

Let $\mathcal{C}$ denote the common items set between test forms X and Y. The A coefficient is calculated for the mean-geometric mean method as

$$A = \prod_{j \in \mathcal{C}} \left(\frac{a_{xj}}{a_{yj}} \right)^{1/n_a}, \tag{7.9}$$

where n_a is the number of common items. For all these three methods, the parameter B is calculated as

$$B = \mu_{b_y} - A\mu_{b_x}. \tag{7.10}$$

Two other methods for obtaining the IRT linking coefficients were proposed by Haebara (1980) and Stocking and Lord (1983). Instead of mean and standard deviations, these methods are based on the item characteristic curve (ICC) and search iteratively for optimal A and B values. In the Haebara method, this is done by minimizing

$$\text{Hcrit} = \sum_i \sum_{j \in \mathcal{C}} \left[p_{ij}(\theta_{yi}, \hat{a}_{yj}, \hat{b}_{yj}, \hat{\gamma}_{yj}) - p_{ij}\left(\theta_{yi}, \frac{\hat{a}_{xj}}{A}, A\hat{b}_{xj} + B, \hat{\gamma}_{xj}\right) \right]^2. \tag{7.11}$$

The Stocking-Lord method instead minimizes the following expression:

$$\text{SLcrit} = \sum_i \left[\sum_{j \in \mathcal{C}} p_{ij}(\theta_{yi}, \hat{a}_{yj}, \hat{b}_{yj}, \hat{\gamma}_{yj}) - \sum_{j \in \mathcal{C}} p_{ij}\left(\theta_{yi}, \frac{\hat{a}_{xj}}{A}, A\hat{b}_{xj} + B, \hat{\gamma}_{xj}\right) \right]^2. \tag{7.12}$$

The ICC-based methods are more robust than the methods based on only the means or means and variances of the distributions of the common item scores in the two samples. Note that the Stocking and Lord method is the most widely applied characteristic curve method (Keller and Hambleton, 2013; Stocking and Lord, 1983).

7.6.1 Concurrent Calibration

Today, when the technology advances allow for calibrating large data sets, most modern programs are moving to concurrent calibration methods for the NEAT designs. These methodologies will require some constraints on the model, such as defining the two groups and allowing for the mean and maybe the variance of θ to be estimated, or making decisions about the item parameters of the common items. For more details, refer to von Davier and von Davier (2007). The use of concurrent calibration within the GKE framework has not yet been explored and is thus left as a topic for future research.

7.7 Equating Transformations for the Local Equating Framework

So far, marginal score distributions $F_X(x)$ and $F_Y(y)$ have been considered when computing the equating transformation. A different approach considers conditional score distributions $F_{X|\lambda}(x \mid \lambda)$ and $F_{Y|\lambda}(y \mid \lambda)$, which was described in Section 4.3.5, leading to a *family* of equating transformations of the form

$$\Psi = \{\varphi : \mathcal{X} \mapsto \mathcal{Y}, \varphi(x; \lambda) = F_{Y|\lambda}^{-1}(F_{X|\lambda}(x))\}, \quad \lambda \in \Lambda, \qquad (7.13)$$

where λ indexes the individual members of the family Ψ, and Λ is the set of possible values for λ (van der Linden, 2000, 2013). Each member of Ψ is known as a local equating transformation (van der Linden, 2011). As λ is usually unknown, an estimate of this parameter is needed in order to use a member of the family to perform the equating.

The local equating method was originally motivated by Lord's equity principle (Lord, 1980). In such a case, λ is replaced by the ability parameter θ, leading to what is called local IRT observed-score equating (OSE) (van der Linden, 2000, 2006). However, λ can also be replaced by a proxy for ability, such as anchor scores or the realized observed score $X = x$ (Wiberg and van der Linden, 2011).

The conditional score probabilities are cumulated and regular equipercentile equating can be used after a continuization method is implemented. The most common continuization method is to use the uniform kernel, also known as linear interpolation. The main difference between IRT OSE and local IRT OSE is that, in the latter, the marginalization on θ is omitted.

The local observed-score KE method (Wiberg et al., 2014) combines IRT local equating and KE. The main difference between local IRT OSE and local IRT observed-score KE is that, in the latter, all the five steps of KE are applied in the equating process. In particular, instead of linear interpolation, kernel continuization is used to obtain continuous approximations of the score probability distributions. The local observed-score KE transformations thus are defined as

$$\varphi(x; \lambda, \boldsymbol{r}(\lambda), \boldsymbol{s}(\lambda)) = F_{Y|\lambda}^{-1}(F_{X|\lambda}(x; \boldsymbol{r}(\lambda)); \boldsymbol{s}(\lambda)), \quad \lambda \in \Lambda, \qquad (7.14)$$

where

$$F_{X|\lambda}(x \mid \lambda) = \Pr(X \leq x \mid \lambda) = \sum_{j, x_j \leq x} r_j(\lambda), \qquad (7.15)$$

$$F_{Y|\lambda}(y \mid \lambda) = \Pr(Y \leq y \mid \lambda) = \sum_{k, y_k \leq y} s_k(\lambda), \qquad (7.16)$$

and $r_j(\lambda) = \Pr(X = x_j \mid \lambda)$ and $s_k = \Pr(Y = y_k \mid \lambda)$ are the conditional score probabilities that define the distributions.

7.8 Graphical Representation of the Equating Transformation

For most versions of the equating transformation functions described in this chapter, the simplest way to obtain a graphical representation of $\varphi(x)$ is by plotting the scores composing the score scale $\mathcal{X}$ against the actual equated values $\hat{\varphi}(x)$. An example of such a plot is shown in Figures 9.4 and 9.5 for equating performed under the EG design.

The situation is a bit different for the equating function obtained under the local equating framework, as in this case, a family of transformations, rather than only one transformation, is defined for different values of λ (see Eq. (7.14)). Thus, a plot of $\varphi(x) = F_{Y|\lambda}^{-1}(F_{X|\lambda}(x))$ will be seen as different sections, each corresponding to a particular value of λ. An example of this type of plot can be seen in Figure 6.3 in González and Wiberg (2017).

7.9 Summary

This chapter has described the equating transformation, which is the fourth step of the GKE framework. The equating transformation computed under the GKE framework was shown to be flexible enough to produce different shapes, such as linear and chained versions of the equating transformation. Moreover, it was also flexible enough to incorporate the IRT framework.

Finally, the concept of local KE transformations, which use conditional score distributions rather than marginal score distributions, was introduced as a natural extension of the GKE framework. The flexibility and adaptability of the GKE framework makes it a powerful tool for addressing a wide range of equating challenges, allowing for tailored solutions based on the specific characteristics and requirements of the data and testing instruments at hand.

8

Evaluating the Equating Transformation

In this chapter, we describe a large number of methods that can be used for evaluating the equating transformation. The chapter is divided into three sections: equating-specific evaluation measures, statistical evaluation measures, and simulation of test scores.

Throughout this chapter, we use *equating-specific* evaluation measures as those which are typically only used and developed for specific equating methods. Further, we refer to *statistical* evaluation measures for those which are used in general statistical inference when evaluating an estimator.

Although there are many equating-specific measures, the focus in the first section is on measures that are suitable for the GKE framework. Similarly, there is a large number of statistical evaluation measures and, in the second section, we will concentrate on those we find most useful when evaluating the equating transformation in the GKE framework.

In the third and final section, we will describe how test scores can be generated to obtain simulated data that are suitable to use when evaluating different aspects of GKE. This section also contains several evaluation measures that require simulated data in order to evaluate the equating transformation.

8.1 Equating-Specific Measures

Equating-specific measures are those that have been developed especially to evaluate equating procedures and are most likely not suitable to use when evaluating statistical estimators in other contexts. Different equating methods have typically been evaluated using different equating criteria, leading to a vast number of equating-specific measures. One reason for the large number of equating-specific measures is that some are only relevant to specific equating methods, and these measures may not be necessary or relevant for all existing equating methods. Using similar evaluation measures with past studies may facilitate the comparison of results between new and old studies. Thus, which measures are chosen when conducting an evaluation may also depend on what has previously been used to evaluate the equating method of interest.

For early reviews on equating evaluation criteria, refer to Han et al. (1997), Harris and Crouse (1993), and Tong and Kolen (2005). These researchers compared different criteria for different equating methods. Some of the equating-specific measures have been used with several equating methods, and in this section, we will give brief descriptions of those which we believe are the most relevant for the GKE framework.

8.1.1 Difference That Matters

The difference that matters (DTM, Dorans and Feigenbaum, 1994) was originally defined as the differences between equated scores and scale scores that are larger than half of a reported score unit. The DTM, or different modifications of it, has been used in a number of equating studies. It has been used in KE in particular by, for example, Andersson and Wiberg (2017) and Wallin and Wiberg (2019). The DTM can be used as a criterion to choose an equating estimator by selecting the one that results in equated values with the smallest number of DTM.

In a broader perspective, the DTM is similar to practical significance, which differs from statistical significance. Statistical significance determines whether a result is likely due to chance or sampling variability, whereas practical significance evaluates whether the result is useful in real-world applications (Kirk, 1996), such as for the test taker.

8.1.2 Percent Relative Error

The percent relative error (PRE, von Davier et al., 2004; Jiang et al., 2012) measures how different the moments of the distribution of the equated values are with respect to those of the score distribution that is defined on the scale to which the test forms are being equated. If the PRE values are high, this indicate a less effective equating (Jiang et al., 2012). The PRE in the $\mathcal{B}$th moment, denoted by PRE $(\mathcal{B})$, is defined as

$$\text{PRE}(\mathcal{B}) = 100 \frac{\mu_{\mathcal{B}}(\varphi(X)) - \mu_{\mathcal{B}}(Y)}{\mu_{\mathcal{B}}(Y)}, \tag{8.1}$$

where $\mu_{\mathcal{B}}(\varphi(X)) = \sum_j (\varphi(x_j))^{\mathcal{B}} r_j$ and $\mu_{\mathcal{B}}(Y) = \sum_k (y_k)^{\mathcal{B}} s_k$. Examples of the use of PRE are given in Chapter 9 for the EG design and Chapter 10 for the NEAT design.

8.1.3 First-Order and Second-Order Equity

As explained in Section 1.1, the *equity* equating requirement establishes that it should be a matter of indifference to a test taker which of the test forms she or he receives. In practice, this means that once equating has been conducted, the conditional distributions of observed (Y) and equated $(\varphi(X))$ scores, at every ability level, should be the same (Lord, 1980). First-order equity (FOE, Morris, 1982) and second-order equity (SOE) are less restrictive versions of

Lord's (1980) equity principle of equating. The former requires that only the conditional means of observed equated scores are equal, whereas in the latter the conditional standard deviations are assumed to be the same. Although test-theory models based on both true scores and ability have been used to evaluate FOE and SOE (Kolen et al., 1992, 1996; Tong and Kolen, 2005), in what follows we assume that an IRT model has been used so that expectations and standard deviations are calculated conditional on IRT ability.

For a binary-scored test form with n items, we define the conditional expected equated score as

$$\psi(\theta) = \mathrm{E}[\varphi(X)|\theta] = \sum_{x=0}^{n} \varphi(x)f(X = x|\theta), \tag{8.2}$$

where $f(X = x|\theta)$ is the conditional distribution of observed scores X equal to x given the ability θ. Thus, FOE holds if $\mathrm{E}[\varphi(X)|\theta] = \mathrm{E}[Y|\theta]$, $\forall \theta$.

The conditional standard error of measurement of $\varphi(X)$ for each ability level is defined as

$$\sigma(\varphi(X)|\theta) = \sqrt{\mathrm{E}[[\varphi(X) - \psi(\theta)]^2|\theta]} = \sqrt{\sum_{x=0}^{n}[\varphi(x) - \psi(\theta)]^2 f(X = x|\theta)}. \tag{8.3}$$

Thus, SOE holds if $\sigma(\varphi(X)|\theta) = \sigma(Y|\theta)$, $\forall \theta$.

An equating transformation can then be evaluated in terms of departures from FOE and SOE using the following measures

$$\mathrm{FOE}_{Dev}(\theta) = \mathrm{E}[\varphi(X)|\theta] - \mathrm{E}[Y|\theta], \tag{8.4}$$

and

$$\mathrm{SOE}_{Dev}(\theta) = \sigma(\varphi(X)|\theta) - \sigma(Y|\theta) \tag{8.5}$$

An example of the use of FOE and SOE within KE is given in Lee et al. (2010).

8.1.4 Standard Error of Equating

Standard error of equating (SEE) can be viewed as a statistical measure, although we have categorized it here as an equating-specific measure. The reason for this is that SEE was first introduced and proposed for use in KE by Holland et al. (1989) (see also von Davier et al., 2004) in order to measure the uncertainty in the estimated equating transformation. Using the KE framework, we have a simple way to obtain the analytical SEE. This is not necessarily the case when using other equating frameworks. We thus decided to place only the more general standard error (SE), described in Section 8.2.4, in the statistical measures group and place the SEE here in the equating-specific measures, as it is only used as a measure of the uncertainty of the equating transformation.

A general definition of the SEE for equating X to Y is

$$\text{SEE}(x) = \hat{\sigma}(x) = \sqrt{\text{Var}(\hat{\varphi}(x))} = \sqrt{\text{Var}(\varphi(x; \hat{r}, \hat{s}))}. \qquad (8.6)$$

To obtain SEE, we thus need the variance of the equating transformation. Note that the SEE is a component-wise function, which means that it can be divided into different parts in order to calculate it. A common way to calculate the SEE is to use the delta method, as described by von Davier et al. (2004), which is briefly described in the following subsection. Another way to calculate the SEE is to use the Bahadur representation of sample quantiles described in Section 8.1.4.2. An example of estimating the SEE under the NEAT design is given in Section 10.8.2.

8.1.4.1   The Delta Method to Obtain the SEE

One possibility for estimating the SEE is to use the well-known δ-method (Bishop et al., 1975; Lehmann, 1999; Rao, 1973; Kendall and Stuart, 1997). The δ-method when estimating the SEE in KE using log-linear models has been described in detail and has been used in von Davier et al. (2004). The descriptions and formulas in this section are based on von Davier et al. (2004), and to facilitate for the readers we are using the same notation. In this section, we provide a brief description of the most important details of this method. For more details, please refer to von Davier et al. (2004). We will start with a general definition of the δ-method as used in KE, and then will give more details for the familiar case of using log-linear models for presmoothing. A difference from von Davier et al. (2004), is that we will also provide a brief description of how to obtain the SEE when IRT models are used in the presmoothing step. The section ends with an example of using the delta method to obtain SEE under a CE NEC design when using propensity scores.

The main theorem states that if $\hat{r}$ and $\hat{s}$ are approximately normally distributed as

$$\mathcal{N}\left(\begin{pmatrix} r \\ s \end{pmatrix}, \Sigma_{\hat{r}, \hat{s}}\right), \qquad (8.7)$$

then for each x, $\hat{\varphi}(x) = \varphi(x; \hat{r}, \hat{s})$ is also approximately normally distributed as

$$\mathcal{N}\left(\varphi(x; r, s), \mathbf{J}_\varphi \Sigma_{\hat{r}, \hat{s}} \mathbf{J}_\varphi^t\right), \qquad (8.8)$$

where $\mathbf{J}_\varphi$ is the Jacobian matrix of the vector-valued function $\varphi(x; r, s)$. Note that when $\hat{r}$ and $\hat{s}$ have been obtained from the fitting of a parametric statistical model $\mathcal{M}(\theta)$ and through the design function, then $\varphi(x; \hat{r}, \hat{s}) = \varphi(x; \mathbf{DF}[\hat{p}(\hat{\theta})]) = F_{h_Y}^{-1}(F_{h_X}(x; \hat{r}(\mathbf{DF}[\hat{p}(\hat{\theta})]), \hat{s}(\mathbf{DF}[\hat{p}(\hat{\theta})]))$. In such cases, the chain rule of derivatives should be applied to find $\mathbf{J}_\varphi$. When the DF is not used to obtain $\hat{r}$ and $\hat{s}$, then it still can happen that the estimated score probabilities are found as a function of other parameters, namely, $\hat{r}(\hat{\theta})$ and $\hat{s}(\hat{\theta})$, and the chain rule of derivatives should still be used to obtain $\mathbf{J}_\varphi$. In both cases, the first term when applying the chain rule of derivatives is calculated as follows.

Let $\partial\varphi/\partial r$ and $\partial\varphi/\partial s$ be the derivatives of the equating transformation with respect to r and s, respectively. Note that $\partial\varphi/\partial r$ denotes the J dimensional row vector whose entries are the J derivatives $\partial\varphi(x;r)/\partial r$ and, similarly, $\partial\varphi/\partial s$ denotes the K dimensional row vector whose entries are the K derivatives $\partial\varphi(x;s)/\partial s$. Using these definitions, the Jacobian of the equating transformation $\mathbf{J}_\varphi$ with dimension $1 \times (J+K)$ is

$$\mathbf{J}_\varphi = \left(\frac{\partial\varphi}{\partial r}, \frac{\partial\varphi}{\partial s} \right). \tag{8.9}$$

The derivatives of the equating transformation can be found using implicit differentiation and the chain rule of derivatives. This results in the following expression:

$$\mathbf{J}_\varphi = \left(\frac{\partial\varphi}{\partial r}, \frac{\partial\varphi}{\partial s} \right) = \frac{1}{F'_Y} \left(\frac{\partial F_X}{\partial r}, -\frac{\partial F_Y}{\partial s} \right), \tag{8.10}$$

where F'_Y represent the derivative of F_Y, i.e., the density f_Y.

The next step is to evaluate $\partial F_X/\partial r$ and $\partial F_Y/\partial s$ in the KE context. This is done by replacing $F_X = F_{h_X}$ and $F_Y = F_{h_Y}$ as defined in Eq. (5.29). Thus, we can instead search for the derivatives of the smoothly continuized CDFs, F_{h_X} and F_{h_Y}, with respect to r and s, respectively. Note that the derivative of $F_{h_X}(x)$ is equal to the density f_{h_X}, which we can write as $F'_{h_X}(x) = f_{h_X}$. Explicit expressions of these terms are found in von Davier et al. (2004). Next we give the details when the derivation of the SEE involves parameters $\hat{r}$ and $\hat{s}$ that have been estimated using DFs and log-linear models.

Let $\boldsymbol{R}$ and $\boldsymbol{S}$ be vectors of presmoothed score distributions from a log-linear model. Recall from Chapter 4 that $\boldsymbol{P}$ contains the probabilities that define the score distributions in population $\mathcal{P}$, and $\mathrm{vec}(\hat{\boldsymbol{P}})$ is the vectorized version of $\boldsymbol{P}$. Similar definitions apply for the probabilities defining the score distributions in population $\mathcal{Q}$ (i.e., $\mathrm{vec}(\hat{\boldsymbol{Q}})$, and $\boldsymbol{Q}$, respectively).

Depending on the data collection design, $\boldsymbol{R}$ and $\boldsymbol{S}$ have different appearances, as can be seen in the first two columns of Table 8.1. Note that for the SG design, we make the convention that the target population $\mathcal{T} = \mathcal{P}$ (see Figure 1.1 and Section 4.1.2). This means that $\hat{\boldsymbol{R}}$ corresponds to $\mathrm{vec}(\hat{\boldsymbol{P}})$, and $\hat{\boldsymbol{S}}$ is an arbitrary vector not influencing either r or s. This is why a hyphen symbol "-" appears in the corresponding entry in Table 8.1. Also note that although the NEAT and NEC designs appear to be the same, their score probabilities depend either on the anchor test scores (NEAT design) or covariates (NEC design). For details on how these expressions were obtained, refer to Chapter 4. The information in the remaining four columns of Table 8.1 will be used later in this subsection.

When the score probability parameters have been obtained using DFs, the chain rule of derivatives is applied, and the Jacobian matrix of the design

TABLE 8.1: The estimated score probability vectors $\boldsymbol{R}$ and $\boldsymbol{S}$, the dimensions of $\boldsymbol{R}$ (J_X) and $\boldsymbol{S}$ (K_Y), and the number of estimated parameters in $\hat{\boldsymbol{R}}$ (T_X), and $\hat{\boldsymbol{S}}$ (T_Y) for five data collection designs.

	$\hat{\boldsymbol{R}}$	$\hat{\boldsymbol{S}}$	J_X	K_Y	T_X	T_Y
EG	$\hat{\boldsymbol{r}}$	$\hat{\boldsymbol{s}}$	J	K	T_r	T_s
SG	$\text{vec}(\hat{\boldsymbol{P}})$	-	JK	-	T_r	-
CB	$\text{vec}(\hat{\boldsymbol{P}}_{12})$	$\text{vec}(\hat{\boldsymbol{P}}_{21})$	JK	JK	$T_{(12)}$	$T_{(21)}$
NEAT	$\text{vec}(\hat{\boldsymbol{P}})$	$\text{vec}(\hat{\boldsymbol{Q}})$	JL	KL	T_P	T_Q
NEC	$\text{vec}(\hat{\boldsymbol{P}})$	$\text{vec}(\hat{\boldsymbol{Q}})$	JL	KL	T_P	T_Q

function $\mathbf{J}_{\text{DF}}$ is included for the estimation of the SEE. Following the general notation, this Jacobian is defined as

$$\mathbf{J}_{\text{DF}} = \begin{pmatrix} \frac{\partial \boldsymbol{r}}{\partial \boldsymbol{R}} & \frac{\partial \boldsymbol{r}}{\partial \boldsymbol{S}} \\ \frac{\partial \boldsymbol{s}}{\partial \boldsymbol{R}} & \frac{\partial \boldsymbol{s}}{\partial \boldsymbol{S}} \end{pmatrix}.$$

As noted in Section 4.1, the mathematical form of the DF depends on the data collection design, and thus its Jacobian will look different depending on the used data collection design. The entries of $\mathbf{J}_{\text{DF}}$ for the different data collection designs are shown in Table 8.2, where $\boldsymbol{I}_J$ and $\boldsymbol{I}_K$ are identity matrices of size $J \times J$ and $K \times K$, $\mathbf{1}_J$, and $\mathbf{1}_K$ are column vectors of ones with dimension J and K, respectively. Note that, as the general formula entries are the same for the NEAT PSE and the NEC PSE designs, they are stated as the same in Table 8.2. Note that to emphasize that we are using covariates instead of anchor scores, we should exchange l for d for the NEC design. Note also that for the EG, SG, and CB designs, the entries in $\mathbf{J}_{\text{DF}}$ are exactly the same as the ones for the Δ matrices that are used to define the DFs for these designs, which is because the DF for these designs is linear (see Section 4.1).

The other component of the SEE, when using the delta method, is the asymptotic covariance matrix of the presmoothed frequencies. If no presmoothing is performed, the covariance matrix of the sample score frequencies, which are assumed to follow a multinomial distribution (i.e., Holland and Thayer, 1987, 2000), can directly be used.

Assume that the presmoothed data $\boldsymbol{R}$ and $\boldsymbol{S}$ are estimated independently, so that $\text{Cov}(\hat{\boldsymbol{R}}, \hat{\boldsymbol{S}}) = 0$. When log-linear models have been used in the presmoothing step, we use so-called "C-matrices" (von Davier et al., 2004; Holland and Thayer, 1987), to estimate the covariance matrices. If $\hat{\boldsymbol{R}}$ and $\hat{\boldsymbol{S}}$ are maximum likelihood estimates of the log-linear models for $\boldsymbol{R}$ and $\boldsymbol{S}$, then the estimated covariance matrix can be calculated as

$$\Sigma_{\hat{R}} = \text{Cov}(\hat{\boldsymbol{R}}) = \mathbf{C}_R \mathbf{C}_R^t, \tag{8.11}$$

$$\Sigma_{\hat{S}} = \text{Cov}(\hat{\boldsymbol{S}}) = \mathbf{C}_S \mathbf{C}_S^t, \tag{8.12}$$

TABLE 8.2: The entries of $\mathbf{J}_{\mathrm{DF}}$ for five different data collection designs.

	EG	SG	CB
$\frac{\partial \boldsymbol{r}}{\partial \boldsymbol{R}}$	$\boldsymbol{I}_J$	$\mathbf{M}$	$w_X\mathbf{M}$
$\frac{\partial \boldsymbol{s}}{\partial \boldsymbol{R}}$	0	$\mathbf{N}$	$(1-w_Y)\mathbf{N}$
$\frac{\partial \boldsymbol{r}}{\partial \boldsymbol{S}}$	0	0	$(1-w_X)\mathbf{M}$
$\frac{\partial \boldsymbol{s}}{\partial \boldsymbol{S}}$	$\boldsymbol{I}_K$	0	$w_Y\mathbf{N}$

	NEAT PSE and NEC PSE
$\frac{\partial \boldsymbol{r}}{\partial \boldsymbol{p}_l}$	$w_{lP}\boldsymbol{I}_J - (1-w)(t_{Ql}/t_{Pl})((t_{Pl})^{-1}\mathbf{p}_l)\mathbf{1}_J^t$
$\frac{\partial \boldsymbol{s}}{\partial \boldsymbol{p}_l}$	$w((t_{Ql})^{-1}\mathbf{q}_l)\mathbf{1}_J^t$
$\frac{\partial \boldsymbol{r}}{\partial \boldsymbol{q}_l}$	$(1-w)((t_{Pl})^{-1}\mathbf{p}_l)\mathbf{1}_K^t$
$\frac{\partial \boldsymbol{s}}{\partial \boldsymbol{q}_l}$	$w_{lQ}\boldsymbol{I}_K - w(t_{Pl}/t_{Ql})((t_{Ql})^{-1}\mathbf{q}_l)\mathbf{1}_K^t$

where $\mathbf{C}_R$ is the $J \times T_r$ matrix $\mathbf{C}_R = N_x^{-1/2}\mathbf{D}_{\sqrt{r}_j}\mathbf{Q}$. In this case, the diagonal matrix $\mathbf{D}_{\sqrt{r}}$ has the entries $\sqrt{r}_j$ along its main diagonal. $\mathbf{Q}$ is the $J \times T_r$ orthogonal matrix from the following $\mathbf{QR}$-factorization $[\mathbf{D}_{\sqrt{r}} - \sqrt{\hat{r}}\hat{r}^t]\mathbf{B}^t = \mathbf{QR}$, where $\mathbf{R}$ is a $T_r \times T_r$ upper triangular matrix and $\mathbf{B}$ is the design matrix of known constants from the log-linear model as defined in Eq. (3.2). Note that the covariance for $\hat{\boldsymbol{S}}$ can be obtained in a similar way. For details on how to obtain these matrices, refer to Holland and Thayer (1987, p.18) and Theorem 3.1 in von Davier et al. (2004).

For instance, for the EG design, we have that

$$\mathrm{Cov}\left(\begin{array}{c}\hat{\boldsymbol{R}} \\ \hat{\boldsymbol{S}}\end{array}\right) = \mathbf{C}\mathbf{C}^t, \tag{8.13}$$

where

$$\mathbf{C} = \begin{pmatrix} \mathbf{C}_R & 0 \\ 0 & \mathbf{C}_S \end{pmatrix}. \tag{8.14}$$

Note that the C-matrices have different appearances depending on the data collection design used, and for a description of them for the SG, CB, and NEAT designs, refer to von Davier et al. (2004) and Holland and Thayer (1987).

When we have obtained the three described parts for calculating the SEE, we use Theorem 5.1 from von Davier et al. (2004). This theorem states that if $\hat{R}$ and $\hat{S}$ are approximately normally distributed as

$$\mathcal{N}\left(\begin{pmatrix} R \\ S \end{pmatrix}, \Sigma_{\hat{R},\hat{S}}\right). \tag{8.15}$$

where

$$\Sigma_{\hat{R},\hat{S}} = \mathrm{Cov}\begin{pmatrix} \hat{R} \\ \hat{S} \end{pmatrix} = \mathbf{C}\mathbf{C}^t,$$

then for each x, $\hat{\varphi}_Y(x) = \varphi_Y(x; \hat{r}, \hat{s})$, is also approximately normally distributed as

$$\mathcal{N}\left(\varphi_Y(x; r, s), \mathbf{J}_{\varphi_Y}\mathbf{J}_{\mathrm{DF}}\mathbf{C}\mathbf{C}^t\mathbf{J}_{\mathrm{DF}}^t\mathbf{J}_{\varphi_Y}^t\right), \tag{8.16}$$

where $\mathbf{J}_{\varphi_Y}$ and $\mathbf{J}_{\mathrm{DF}}$ are the Jacobian matrices described earlier and $\mathbf{C}$ is the factor of the covariance matrix of $\hat{R}$ and $\hat{S}$ described in Eq. (8.13). Thus, the SEE can be calculated as

$$\mathrm{SEE}\,(x) = \|\hat{\mathbf{J}}_\varphi\hat{\mathbf{J}}_{\mathrm{DF}}\mathbf{C}\|, \tag{8.17}$$

where $\|\cdot\|$ is the Euclidean norm. Note that Eq.(8.16) is a particular case of Eq. (8.8) when we use log-linear models in the presmoothing step. For more details on how to calculate the SEE in KE using the delta method with log-linear models for different designs, refer to von Davier et al. (2004).

When IRT models have been used instead of log-linear models in the presmoothing step, the $\hat{R}$ and $\hat{S}$ are not obtained through DFs, but rather using one of the methods described in Section 4.3, most usually, the Lord-Wingersky algorithm. Because the estimated score probabilities are functions of the item parameters, the uncertainty comes from the estimation of the latter, and thus their covariances are included in the calculation of the SEE. More specifically, suppose that an IRT model with item parameter vector ω has been fitted to the data using maximum likelihood (Bock and Aitkin, 1981; Lord, 1980). Under some regularity conditions, $\hat{\omega}$ is asymptotically normally distributed with covariance $\Sigma_{\hat{\omega}}$ (Ogasawara, 2003). Then, the estimation of the score probabilities, $\hat{R}(\theta, \hat{\omega})$ also follows approximately a normal distribution with covariance $\Sigma_{\hat{R}}$, where

$$\Sigma_{\hat{R}} = \frac{\partial R(\theta, \omega)}{\partial \omega}\Sigma_{\hat{\omega}}\frac{\partial R(\theta, \omega)}{\partial \omega}^t.$$

A similar reasoning applies to $\hat{S}$. Thus, these covariances are used in (8.8). Note that the asymptotic covariances will have different appearances for different designs, but they are needed to analytically calculate the SEE in a similar way as above when log-linear models have been used. For a detailed description of how to obtain SEE with the delta method when IRT models have been used in the presmoothing step with the NEAT design, refer to Andersson and Wiberg (2017).

If discrete associate kernels or the beta4 models have been used in the presmoothing step, covariances of $\hat{\boldsymbol{R}}$ and $\hat{\boldsymbol{S}}$ have not yet been derived for all designs, and this is thus a good topic for future research.

Recently, Marcq and Andersson (2022) proposed a modified method for calculating the SEE that also accounts for the bandwidth estimation variability when using the penalty method (see Section 6.3 and Eq. (6.7)). These authors, however, focused only on the first out of two penalty terms, i.e., PEN1, arguing on the complexity of accounting for estimation variability with the second penalty term, PEN2. As a matter of fact, a problem with Eq. (6.7) is that the PEN2 part is not differentiable. This means that we cannot obtain the Jacobian of the PEN function and thus not use Eq. (8.17). A solution might be to use optimization methods that do not rely on differentiation of the objective functions. The family of *search* methods of optimization (e.g., Hooke and Jeeves, 1961; Du et al., 2016) might then be a valuable alternative for this type of optimization. Although we do not expect to see any large differences with the use of a fully differentiable penalty function, we leave the development of such a function as a possible topic for future research.

In von Davier et al. (2004), the SEE for the EG, SG, CB, and NEAT designs are given when log-linear models were used in the presmoothing step. We thus chose here to illustrate how to obtain the SEE with the NEC design, which was not described in that book. In particular, we describe how to use the delta method to obtain the SEE for the CE NEC design when using propensity scores (PS), which was first proposed by Wallin and Wiberg (2019) and described in Section 4.1.5.

We start by defining the propensity score for strata d of test forms X and Y as e_{Xd} and e_{Yd}, respectively. In the NEC CE design, we first link X to $e_{Xd}(\mathbf{D})$ in population P, and then $e_{Yd}(\mathbf{D})$ to Y in population Q. In order to do this, we assume that these links are population invariant for a population $\mathcal{T}$. We start from the common approach for the NEAT design using the CE method, although stratified propensity scores are used instead of anchor scores, and view the two links in CE as equating transformations from two SG designs. The Jacobian of the equating transformation, $\mathbf{J}_\varphi(x)$ can thus be formed from two Jacobians and four sets of probabilities: $\boldsymbol{r}_P = (r_{P1}, ..., r_{PJ})^t$, $\boldsymbol{t}_P = (t_{P1}, ..., t_{PD})^t$, $\boldsymbol{t}_Q = (t_{Q1}, ..., t_{QD})^t$, and $\boldsymbol{s}_Q = (s_{Q1}, ..., s_{QK})^t$, where $t_{Pd} = \Pr(e_{Xd}(\mathbf{D}) = e_{Xd}(\mathbf{d})|P)$ and $t_{Qd} = \Pr(e_{Yd}(\mathbf{D}) = e_{Yd}(\mathbf{d})|Q)$. This gives us

$$\mathbf{J}_\varphi(x) = (\varphi'(e_{Yd})\mathbf{J}_{\varphi_{e_{Xd}}}(x), \mathbf{J}_\varphi(e_{Yd})), \tag{8.18}$$

where $\varphi'(e_{Yd})$ is the derivative of $\varphi(e_{Yd})$ with respect to the stratified propensity score in strata d, e_{Yd}, $\mathbf{J}_{\varphi_{e_{Xd}}}(x)$ is the Jacobian of $\varphi(e_{Yd})$, and $\mathbf{J}_\varphi$ is the Jacobian of φ.

Recall how the design function for the NEAT CE design was defined in (4.9). To obtain the SEE for the NEC CE with propensity scores, the two links that are used can be seen as equating transformations from two SG designs, meaning that the Jacobian of the equating transformation can be obtained from two Jacobians and four sets of probabilities (r_P, t_P, s_Q and t_Q).

Assume that $\mathbf{P}$ and $\mathbf{Q}$ are to be estimated independently with log-linear models and using maximum likelihood. In this case, we are not searching for a single $\mathbf{C}$ matrix, but instead we want to find the $\mathbf{V}$ matrices to calculate the covariances of each of the two links as follows:

$$\mathrm{Cov}\begin{pmatrix} \hat{r}_P \\ \hat{t}_P \end{pmatrix} = \mathbf{V_P V_P^t}, \tag{8.19}$$

and

$$\mathrm{Cov}\begin{pmatrix} \hat{s}_Q \\ \hat{t}_Q \end{pmatrix} = \mathbf{V_Q V_Q^t}, \tag{8.20}$$

where

$$\mathbf{V_P} = \begin{pmatrix} \mathbf{M_P} \\ \mathbf{N_P} \end{pmatrix} \mathbf{C_P}, \tag{8.21}$$

and

$$\mathbf{V_Q} = \begin{pmatrix} \mathbf{N_Q} \\ \mathbf{M_Q} \end{pmatrix} \mathbf{C_Q}. \tag{8.22}$$

The SEE for NEC CE design with propensity scores can then be defined as

$$\mathrm{SEE}\,(x) = \left\| \hat{\mathbf{J}}_\varphi \hat{\mathbf{V}} \right\|,$$

where

$$\mathbf{V} = \begin{pmatrix} \mathbf{V_P} & 0 \\ 0 & \mathbf{V_Q} \end{pmatrix}.$$

8.1.4.2   Bahadur Representation to Obtain the SEE

Instead of using the delta method to obtain the SEE, González and Wallin (2021) proposed using a method that they called *quantile-based standard error of equating* (QB-SEE), and which is based on the Bahadur representation of sample quantiles (Bahadur, 1966). The QB-SEE is defined as

$$\mathrm{SEE}_Y^B(x) = \frac{1}{F'_{h_Y}(\varphi)} \left\{ \mathrm{Var}(\hat{F}_{h_X}(x)) + \mathrm{Var}(\hat{F}_{h_Y}(\varphi)) - 2\mathrm{Cov}(\hat{F}_{h_X}(x), \hat{F}_{h_Y}(\varphi)) \right\}^{1/2},$$
$$\tag{8.23}$$

where $\hat{F}_{h_X}(x)$ and $\hat{F}_{h_Y}(y)$ are defined as in (5.29) and (5.30), respectively, and $F'_{h_Y}(\varphi) = \frac{dF_{h_Y}(t)}{dt}$ evaluated at $t = \varphi$. The variances and covariance terms in (8.23) must be calculated according to the specific equating design that is being considered. For instance, for the NEAT design, following Liou et al.

(1997), they are calculated as

$$
\begin{aligned}
\mathrm{Var}(\hat{F}_{h_X}(x)) &= \mathrm{Var}\left(\sum_j \hat{r}_j K(\hat{R}_{jX}(x))\right) \\
&\approx \sum_j \sum_{j'} K\left(R_{jX}(x)\right) K\left(R_{j'X}(x)\right) \mathrm{Cov}\left[\hat{r}_j, \hat{r}_{j'}\right] \\
&= \sum_j K_{jX}^2\left(R_{jX}(x)\right) \mathrm{Var}\left[\hat{r}_j\right] \\
&\quad + \sum_{j \neq j'} K_{jX}\left(R_{jX}(x)\right) K_{j'X}\left(R_{jX}(x)\right) \mathrm{Cov}\left[\hat{r}_j, \hat{r}_{j'}\right],
\end{aligned}
\tag{8.24}
$$

where $R_{jX}(x) = \frac{x - a_{jX}x_j - (1 - a_{jX})\mu_X}{a_{jX}h_{jX}}$, $r_j = \mathrm{Pr}(X = x_j)$, and with

$$
\begin{aligned}
\mathrm{Var}\left[\hat{r}_j\right] = \sum_l &\left\{ \frac{\hat{r}_{j|l}[1 - \hat{r}_{j|l}]}{(N_x + 1)\hat{t}(a_l) - 1}\hat{t}^2(a_l) + \frac{\hat{t}(a_l)[1 - \hat{t}(a_l)]}{N_x + N_y}\hat{r}_{j|l}^2 \right. \\
&\left. + \frac{\hat{r}_{j|l}\left[1 - \hat{r}_{j|l}\right]\hat{t}(a_l)\left[1 - \hat{t}(a_l)\right]}{\left[(N_x + 1)\hat{t}(a_l) - 1\right](N_x + N_y)} \right\} - \sum_{l \neq l'}\sum \frac{\hat{t}(a_l)\hat{t}(a_{l'})}{N_x + N_y}\hat{r}_{j|a}\hat{r}_{j|a'},
\end{aligned}
\tag{8.25}
$$

and

$$
\begin{aligned}
\mathrm{Cov}\left[\hat{r}_j, \hat{r}_{j'}\right] = \sum_a &\left\{ \frac{\hat{t}(a_l)[1 - \hat{t}(a_l)]}{N_x + N_y}\hat{r}_{j|l}\hat{r}_{j'|l} - \frac{\hat{r}_{j|l}\hat{r}_{j'|l}}{(N_x + 1)\hat{t}(a_l) - 1}\hat{t}^2(a_l) \right. \\
&\left. - \frac{\hat{r}_{j|l}\hat{r}_{j'|l}\hat{t}(a_l)[1 - \hat{t}(a_l)]}{\left[(N_x + 1)\hat{t}(a_l) - 1\right](N_x + N_y)} \right\} - \sum_{l \neq l'}\sum \frac{\hat{t}(a_l)\hat{t}(a_{l'})}{N_x + N_y}\hat{r}_{j|l}\hat{r}_{j'|l'},
\end{aligned}
\tag{8.26}
$$

where, $t_l = \mathrm{Pr}(A = a_l)$ are the marginal score probabilities for the anchor random variable A, and $r_{j|l}$ and $s_{k|l}$ are the conditional score probabilities of X given A, and Y given A, respectively. Replacing terms accordingly, similar derivations lead to obtaining the variance of F_Y. Finally, the covariance term is calculated as

$$
\begin{aligned}
\mathrm{Cov}(\hat{F}_{h_X}(x), \hat{F}_{h_Y}(\varphi)) \approx \sum_j \sum_k K(R_{jX}) K(R_{kY}) &\left\{ \sum_l \frac{\hat{t}(a_l)[1 - \hat{t}(a_l)]}{N_x + N_y}\hat{r}_{j|l}\hat{s}_{k|l} \right. \\
&\left. - \sum_{l \neq l'}\sum \frac{\hat{t}(l)\hat{t}(l')}{N_x + N_y}\hat{r}_{j|l}\hat{s}_{k|l'} \right\}.
\end{aligned}
\tag{8.27}
$$

The results obtained using either the QB-SEE or the SEE under the NEAT design are very similar (González and Wallin, 2021), and thus no examples of this approach are given here. An advantage of the QB-SEE method is that it allows for the separation of sources of uncertainty influencing the SEE, as can be grasped from Eq. (8.23). Another advantage is that it does not rely on normality. This clears the way for other models and methods for presmoothing that also do not rely on normality, for example the discrete kernel presmoothing method described in Chapter 3.

8.1.5 Bootstrap Standard Error of Equating

The nonparametric bootstrap method (Efron, 1982; Efron and Tibshirani, 1993) can be used to obtain SEEs by using a large number of random samples with replacement from the (unique) observed sample data. Bootstrap SEEs are useful in equating because it is difficult to derive analytically explicit formulas for the SEEs in some equating methods. To obtain nonparametric bootstrap SEEs, we can use the following algorithm:

1. Draw a random sample with replacement of N_X test scores coming from the observed sample of scores from test form X.

2. Draw a random sample with replacement of N_Y test scores coming from the observed sample of scores from test form Y.

3. Estimate the equating transformation using the data from the bootstrap samples obtained in Steps 1 and 2.

4. Replicate steps 1–3 L times to obtain L bootstrap estimates of the equating transformation.

5. Estimate SEEs at test score x_j using

$$\text{SEE}(x_j) = \sqrt{\frac{1}{L} \sum_{l=1}^{L} [\hat{\varphi}_l(x_j) - \bar{\hat{\varphi}}_l(x_j)]^2}, \qquad (8.28)$$

where $\bar{\hat{\varphi}}_l(x_j) = \frac{1}{L} \sum_{l=1}^{L} \hat{\varphi}_l(x_j)$.

Instead of using nonparametric bootstrap, parametric bootstrap (Efron and Tibshirani, 1993) can also be considered. In parametric bootstrap, a parametric model is fitted to the data, and this model is treated as if it describes the population appropriately. This means that the statistics of interest are computed on samples generated from a statistical model. In our case, we are interested in the equating transformation. However, as it is not possible to directly sample from an equating transformation, one possibility is instead to sample test scores from fitted score distribution described by a statistical model in the presmoothing step. Using these sampled test scores, we can estimate the equating transformation. As the populations are assumed to have infinite size, sampling with or without replacement is thus considered to be the same (Kolen and Brennan, 2014). To obtain parametric bootstrap SEEs, we can use the following algorithm:

1. Use statistical models (for example, log-linear models as described in Section 3.2.1) to fit the score distribution of the test scores X from test form X.

2. Use statistical models (for example, log-linear models as described in Section 3.2.1) to fit the score distribution of the test scores Y from test form Y.

3. Use the fitted distribution from Step 1 as the population distribution for test scores X and randomly select N_X scores from this population distribution. The obtained distribution is the parametric bootstrap sample distribution of scores X.

4. Use the fitted distribution from Step 2 as the population distribution for test scores Y and randomly select N_Y scores from this population distribution. The obtained distribution is the parametric bootstrap sample distribution of scores X.

5. Estimate the equating transformation using the parametric bootstrap sample distributions of scores obtained in Step 3 and Step 4.

6. Repeat Steps 3, 4, and 5 L times.

7. Estimate the SEE at test score x_j using Eq. (8.28).

Note that if IRT equating methods are used, the parametric bootstrap requires that the item responses are generated using IRT models. One example of bootstrap SEE is given in Chapter 9 for the EG design, and another example is given in Chapter 10 for the NEAT design.

8.1.6 Indices to Compare Equating Transformations

Han et al. (1997) used a number of different indices to evaluate the discrepancies between two equating transformations $\hat{\varphi}_{Y_1}$ and $\hat{\varphi}_{Y_2}$. The indices can be used to either compare equating transformations obtained using different equating methods or to compare an estimated equating transformation against a true equating transformation. We postpone the discussion of what we mean with a true equating transformation until Section 8.2. The three first indices described in what follows all give a single value as they are all different averages over the total number of scores. They can be used whenever we want to compare two equating transformations used to equate scales containing the same number of possible scores. The larger the index value, the greater the difference is between the examined equating methods. Values close to zero indicate that the two equating methods give similar equated values. An example of the use of these three indices when equating is conducted under the NEAT design is provided in Section 10.8.5. The section ends with a description of a fourth index, proposed by von Davier et al. (2004), that has been used to compare kernel equating transformations.

8.1.6.1 Mean Signed Difference

The mean signed difference (MSD) between two equating transformations is a negative or a positive number obtained by averaging the comparison of the equated values at each score point. MSD was proposed for use in the evaluation of equating transformations by Han et al. (1997) and is formally defined as

$$\text{MSD} = \frac{1}{n}\sum_{j=1}^{n}[\hat{\varphi}_{Y1}(x_j) - \hat{\varphi}_{Y2}(x_j)]. \tag{8.29}$$

8.1.6.2 Mean Absolute Difference

The mean absolute difference (MAD) was another index appearing in Han et al. (1997). This index is either zero, if there are no differences between the compared equating transformation, or positive as absolute values are used. The MAD is formally defined as

$$\text{MAD} = \frac{1}{n}\sum_{j=1}^{n}|\hat{\varphi}_{Y1}(x_j) - \hat{\varphi}_{Y2}(x_j)|. \tag{8.30}$$

8.1.6.3 Root Mean Squared Difference

The root mean squared difference (RMSD) was also used by Han et al. (1997) to compare equating transformations with each other. This index is zero if there is no difference or positive if there are differences. The RMSD is formally defined as

$$\text{RMSD} = \sqrt{\frac{1}{n}\sum_{j=1}^{n}[\hat{\varphi}_{Y1}(x_j) - \hat{\varphi}_{Y2}(x_j)]^2}. \tag{8.31}$$

8.1.6.4 Standard Error of Equating Difference

The standard error of equating difference (SEED, von Davier et al., 2004) was proposed as a standard error for the differences between two KE transformations. It was developed to be used as a tool for deciding between competing transformations, especially for choosing between the linear equating transformation version obtained for large values of the bandwidth parameter, and the KE transformation that uses an optimally estimated bandwidth. If the difference is small, one could consider using the simpler linear equating transformation. Note that the SEED has also been proposed for use with traditional equating measures (Moses and Zhang, 2011).

To define SEED in GKE, let $\hat{\varphi}_1$ and $\hat{\varphi}_2$ represent two equating transformations. When log-linear models have been used in the presmoothing step, the

SEED is defined as the Euclidean norm of the difference between the following two vectors of the SEEs:

$$\text{SEED}_Y(x) = \|\mathbf{J}_{\varphi_1}\mathbf{J}_{\text{DF}}\mathbf{C} - \mathbf{J}_{\varphi_2}\mathbf{J}_{\text{DF}}\mathbf{C}\| = \sqrt{\text{Var}\left(\hat{\varphi}_1(x) - \hat{\varphi}_2(x)\right)}. \quad (8.32)$$

An example of the use of SEED in the NEAT design is given in Section 10.8.4.

8.2 Statistical Measures

In regular statistical inference, the evaluation of an estimator $\hat{\theta}(\mathbf{X})$ of a parameter θ is made using measures of the average difference between the estimator and the parameter. A common choice is the mean squared error (MSE) which measures the average square difference in the form $\text{E}_\theta(\hat{\theta}(\mathbf{X}) - \theta)^2$. This and other statistical measures of evaluation are based on expectations of an estimator. This is different from the previously described equating-specific measures, which do not focus on estimating a parameter. However, if the equating transformation is viewed as a parameter which we want to estimate, statistical measures could also be defined to evaluate the estimated equating transformation. As expectations are not generally available in closed forms (see, for example, Section 8.2.1), simulated test score data are often used to approximate the expectations and thus calculate the statistical measures. Simulated scores are especially useful when we want to evaluate an equating transformation under different conditions. In Section 8.3, we will describe different options for simulating test score data.

When using statistical evaluation measures, the true and estimated equated values are compared for each test score. If we use simulated data, we can follow the suggestions by Wiberg and González (2016), where an explicit probability model was used to generate score data in order to obtain true equated values. Note that the mathematical form of the equating transformation will change depending on the framework used. As there is no unique true equating model, the true equated scores are obtained from the model that the researcher has chosen as the true model. A researcher must therefore carefully define what is considered to be a true model as there might be several alternatives.

In order to obtain statistical measures with simulated data, we proceed as follows. Let $\hat{\varphi}(x)$ denote an estimate of the true equating transformation $\varphi(x)$. Denote the test scores as x_j, $j = 1, \ldots, J$ and assume we use L replications (or samples). Let $\hat{\varphi}_l(x_j)$ be the equated score based on sample l and let $\varphi(x_j)$ denote the true equated score. We can calculate the different statistical measures approximating the expectations by using a large number of replications, as we will describe in the following subsections.

If, instead of simulations, we want to use statistical evaluation measures with empirical data, we could follow the suggestion of Lord (1980), which was implemented for KE by Leoncio et al. (2023), in which a test form is equated to itself.

In the following subsections, we define a number of useful statistical evaluation measures adopted for the GKE context. Several of these statistical measures will be exemplified in Section 8.3.3.

8.2.1   Bias

The bias of an estimator is the difference between the expected value of the estimator of interest and the true value of the parameter that is being estimated. This means that bias can take both positive and negative numbers. Bias is considered to be a systematic error, and we thus want the bias to be small. If an estimator lacks bias, it is referred to as an unbiased estimator. In the equating context, we can define the bias of the equating transformation as

$$\text{Bias}(\hat{\varphi}) = E_{f_\varphi}(\hat{\varphi} - \varphi). \tag{8.33}$$

where the expectation E_{f_φ} is taken over the distribution of the equating transformation estimates f_φ. In practice, we use replicated data to calculate the bias, as is exemplified in Section 8.3.3. If we use L replications, we can estimate the bias of a specific equated test score x_j, $j = 1, \ldots, J$ from

$$\text{Bias}(x_j) = \frac{1}{L} \sum_{l=1}^{L} [\hat{\varphi}_l(x_j) - \varphi(x_j)]. \tag{8.34}$$

Bias has been used in KE by several researchers. It has, for example, been used for comparing equating transformations obtained from using different kernels in KE (Wiberg and González, 2016), and for comparing equating transformations from IRT OSE and KE using simulated data (Leoncio et al., 2023).

8.2.2   Mean Squared Error

The mean squared error (MSE) of an estimator is the expected value of the squared difference between the estimator and the true value of the parameter that is being estimated. This means that MSE is either zero or positive and a lower value is preferred over higher values. A general definition of the MSE in the equating context is

$$\text{MSE}(\hat{\varphi}) = E_{f_\varphi}[(\hat{\varphi} - \varphi)^2]. \tag{8.35}$$

Similarly to bias, we use replicated data to calculate the MSE, as will be exemplified in Section 8.3.3. If we use L replications, we can estimate the MSE for test score x_j as follows:

$$\text{MSE}(x_j) = \frac{1}{L} \sum_{l=1}^{L} [\hat{\varphi}_l(x_j) - \varphi(x_j)]^2. \tag{8.36}$$

MSE has been used with KE in several research studies. It has, for example, been used when comparing equating transformations with different kernels in KE (Wiberg and González, 2016), when comparing different methods to select bandwidth in KE (Häggström and Wiberg, 2014), and when comparing equating transformations from IRT OSE with KE using simulated data (Leoncio et al., 2023).

8.2.3 Root Mean Squared Error

The root mean squared error (RMSE) is the squared root of the MSE. The RMSE takes zero or positive values and smaller RMSE values are preferable over larger values. The RMSE in the equating context can be formally defined as

$$\text{RMSE}(\hat{\varphi}) = \sqrt{\text{E}_{f_\varphi}[(\hat{\varphi} - \varphi)^2]}, \tag{8.37}$$

which for a specific test score can be estimated with

$$\text{RMSE}(x_j) = \sqrt{\frac{1}{L} \sum_{l=1}^{L} [\hat{\varphi}_l(x_j) - \varphi(x_j)]^2}. \tag{8.38}$$

RMSE has been used in several KE research studies in the past. For example, it has been used to evaluate different equating transformations using different kernels (Wiberg and González, 2016), and when examining the use of propensity scores in KE (Wallin and Wiberg, 2019). An example of the use of RMSE (with Gaussian and uniform kernels) is provided in Section 8.3.3, and the next subsection describes how the RMSE can be divided into systematic and random errors.

8.2.4 Standard Error

The standard error (SE) of a parameter estimate is the standard deviation of the sampling distribution of the parameter. The SE is a random error, and if used in equating it can provide a measure of the uncertainty of the equating estimates. To obtain the sampling distribution of the equating transformation we can proceed as follows. Obtain random independent samples. For each sample, calculate the equating transformation. After several repetitions we obtain a distribution of the equating transformation, which has a variance and mean on its own. If we increase the sample size, the equating transformation becomes closer to the population equating transformation.

The previous defined RMSE contains both systematic errors (bias) and random errors (SE). The SE can be obtained if we have the RMSE and bias, as we can define the RMSE in test score x_j as

$$\text{RMSE}(x_j) = \sqrt{\text{Bias}(x_j)^2 + \text{SE}(x_j)^2}.$$

This means that we can obtain the SE if we have the RMSE and bias, an approach used by Wiberg and González (2016). Note that we can also obtain the SE if we have the MSE and bias, as we can define the MSE in test score x_j as

$$\text{MSE}(x_j) = \text{Bias}(x_j)^2 + \text{SE}(x_j)^2.$$

SE in the context of equating transformations can be formally defined as follows:

$$\text{SE}(\hat{\varphi}) = \sqrt{\text{Var}(\hat{\varphi}(x))}. \tag{8.39}$$

Note that in the context of equating, the definition of SE coincides with the definition of the SEE. SE can be used to emphasize that replications have been used to obtain the standard errors, and SEE can similarly be used to emphasize that they have been obtained analytically. In practice, the SE can sometimes be tricky to obtain analytically, and thus it is common to use bootstrap SEs as described for bootstrap SEEs in Section 8.1.5. An example of the use of bootstrap SEs when comparing bandwidth methods in KE is provided in Wallin et al. (2021).

8.2.5   Cumulants

In GKE, we are often interested in comparing how similar or different distributions are. Moments and cumulants were described in Section 5.5, and cumulants were mentioned as a possible evaluation tool, in addition to moments, to determine how different or similar the distributions $X(h_X)$ and X are. As mentioned in that section, the first cumulant corresponds to the mean, the second cumulant to the variance, and the third cumulant is equivalent to the third central moment (skewness), respectively. However, higher-order cumulants do not directly correspond to higher-order moments. An example of the use of cumulants in KE to compare different $X(h_X)$ distributions when different kernels have been used is given in Lee and von Davier (2011).

8.3   Simulating Test Scores

Simulations are experiments that are used in order to control the environment when examining different conditions of interest. The typical procedure when using simulations is to generate data from a known probability model and then vary different conditions under controlled forms and compare the estimated parameters under the different conditions with the known true parameters. This approach is very useful if we want to evaluate an equating transformation under different conditions.

There are a number of different probability models that can be used as a true model when simulating test scores. In test score equating, several different models have been used, such as log-linear models as described in Section 3.2.1, IRT models as described in Section 3.4.2, and the beta-binomial model as described in Section 3.4.1. González et al. (2016) also recently proposed the use of the Poisson's binomial model, as described in Section 4.3.2.

Regardless of which probability model is chosen as the true model, we need to be careful when designing the study to ensure that none of the equating methods are favored. Sometimes this means using more than one true probability model. For a general discussion on evaluating an equating transformation with simulations, see Wiberg and González (2016) and González and Wiberg (2017), Chapter 7. Also refer to Leoncio et al. (2023) for evaluating and comparing equating transformations in IRT OSE and KE.

The choice of the true model depends on what we are interested in. We may be interested in mimicking the distributions of a real test or some of the real test items. We may also be interested in different shapes of the test score distribution and may thus need to generate symmetric, skewed, or bimodal test score distributions. To mimic a real test, we could use the multinomial distribution to generate test scores from known score probabilities. An example using the multinomial distribution to generate test scores is given in Section 8.3.3.

If we are interested in certain item patterns, we could use IRT models with different distributions for the item parameters. To facilitate comparison between obtained results with other research studies, it may be important to use the same or similar item parameters as in those research studies. This was done, for example, in Andersson and Wiberg (2017), who used the same distributions for the item parameters as in Ogasawara (2003). As an example, both of these studies used the following distributions for the item parameters: $U(0.5, 2)$ for the item discrimination, $\mathcal{N}(0, 1)$ for the item difficulty, and $\mathcal{N}(0.25, 0.05)$ for the item guessing.

We can also use beta-distributed populations to define the true score probabilities of X and Y. The generated data can then be rounded and multiplied by the largest possible score of each respective test form. To get different test distributions from the beta distributions, one could, for instance, change the shape parameters (α and β) as follows. For a symmetric distribution, we can use $\alpha = 5$ and $\beta = 5$; for a negatively skewed distribution, we can use $\alpha = 5$ and $\beta = 2$; for a positively skewed distribution, we can use $\alpha = 2$ and $\beta = 5$; and to obtain a bimodal distribution, we can use a mixture of $\alpha = 25$ and $\beta = 15$, and $\alpha = 15$ and $\beta = 25$, as was used in, for example, Wallin et al. (2021).

It is also possible to use different distributions for the test takers' ability. For example, we could use $\mathcal{N}(0,1)$ or different standardized χ^2 distributions as in Ogasawara (2003) or Andersson and Wiberg (2017).

Once we have decided upon the true model we can move forward and perform a simulation study. Regardless of the chosen true model, we proceed similarly. It is important, however, to be aware that the choice of the generating model could affect the final equating values. A simulation study to examine the estimator of the equating transformation can be conducted by using the following steps:

1. Select one or several true score model(s).

2. From the model(s) in Step 1, obtain true values for the score probabilities and estimate the true equating transformation using the true model(s).

3. Use the known probability model(s) to create simulated score data for the two test versions.

4. Estimate the equating transformations for each generated data set.

5. Repeat Steps 3 and 4 for a large number of replications.

6. Evaluate the equating transformation by comparing the estimated equating transformations with the true equating transformation using statistical measures.

Using simulated data to compare equating transformations within a framework such as KE is relatively straightforward (Wiberg and González, 2016). For example, we might want to compare how different kernels affect the equating transformation. However, it quickly becomes more complicated when comparing equating transformations between frameworks, e.g. KE and IRT OSE, as one has to be careful not to favor any specific methods or framework depending on how the comparison is set up (Leoncio and Wiberg, 2018; Leoncio et al., 2023). For example, to avoid favoring IRT OSE or IRT KE when comparing them with KE, Leoncio et al. (2023) generated test scores using both the beta distribution and an IRT model.

8.3.1 Simulated Equating-Specific Measures

The previously described equating-specific measures were originally developed for empirical data. If we use simulations with replications, we can adjust the previously described equating-specific measures as proposed by Wiberg and González (2016). The SEEs for each score value x_j can be averaged (denoted ave) over L replications as follows:

$$\text{SEE}_Y^{ave}(x_j) = \frac{1}{L}\sum_{l=1}^{L}\text{SEE}_Y^{(l)}(x_j), \tag{8.40}$$

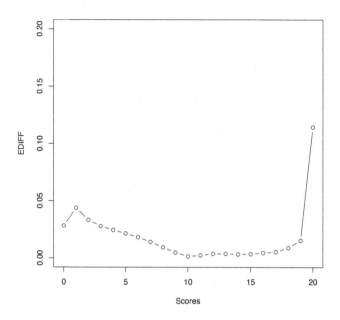

FIGURE 8.1: EDIFF between a Gaussian and uniform kernel.

where $\mathrm{SEE}_Y^{(l)}(x_j)$ is the value of the SEE defined in Eq. (8.6) for the lth replicate.

To adjust PRE to be used with replicated data, PRE can be averaged over replications, and we then get

$$\mathrm{PRE}_{ave}(\mathcal{B}) = \frac{1}{L}\sum_{l=1}^{L}\mathrm{PRE}^{(l)}(\mathcal{B}), \qquad (8.41)$$

where $\mathrm{PRE}^{(l)}(\mathcal{B})$ is the value of the PRE defined in Eq. (8.1) for the lth replicate.

An averaged measure of DTM, which can be used to decide if there are differences between two equating estimators is the equating difference (EDIFF). EDIFF can be defined with replicated data as

$$\mathrm{EDIFF} = \frac{1}{L}\sum_{l=1}^{L}|\hat{\varphi}_{Y_1}^{(l)}(x_j) - \hat{\varphi}_{Y_2}^{(l)}(x_j)|, \qquad (8.42)$$

where $\hat{\varphi}_{Y_1}$ and $\hat{\varphi}_{Y_2}$ represent the estimates from two different equating estimators (Liang and von Davier, 2014). Figure 8.1 shows an example of EDIFF when examining the differences between using uniform and Gaussian kernels, as examined in Section 8.3.3. In this example, the largest differences appear in the lower and the highest scores, although none of the differences are a DTM.

8.3.2 Simulated Comparison Indices

The first three indices for comparing equating transformations in Section 8.1.6 can be modified to be suitable for a simulation study. Wiberg and González (2016) proposed modifications to MSD, MAD, and RMSD in order to use them with a full set of replications. Their modifications became average measures of the comparison indices. The MSD was redefined as the average mean signed difference (AMSD) and defined as

$$\text{AMSD}(\varphi_Y) = \frac{1}{L} \sum_{l=1}^{L} \text{MSD}_l. \tag{8.43}$$

The MAD was redefined as the average mean absolute difference (AMAD) and defined as

$$\text{AMAD}(\varphi_Y) = \frac{1}{L} \sum_{l-1}^{L} \text{MAD}_l. \tag{8.44}$$

The RMSD was redefined as the average root mean squared difference (ARMSD) and defined as

$$\text{ARMSD}(\varphi_Y) = \frac{1}{L} \sum_{l=1}^{L} \text{RMSD}_l. \tag{8.45}$$

These three definitions are in line with how Chen (2012) used these measures, although no explicit formula were provided for them.

As all these measures give a single value, Wiberg and González (2016) proposed modifying AMAD in order to have a measure with values at each score over the full score range. Their proposed measure was the average point absolute difference (APAD), defined as

$$\text{APAD} = \frac{1}{L} \sum_{l=1}^{L} |\hat{\varphi}_Y^{(l)}(x_j) - \varphi_Y(x_j)|, \tag{8.46}$$

Wiberg and González (2016) further noted that if AMSD was similarly modified to average point signed difference (APSD), the expression is equivalent to bias, and if AMSD is modified to average point root squared difference (APRSD), it is mathematically equivalent to APAD. These definitions are thus excluded here. The four simulated comparison indices are included in an example in the next subsection.

8.3.3 Examples of Statistical and Simulated Measures

To illustrate how to obtain some of the described statistical measures in the equating context, we used simulated data with 1,000 replications to evaluate KE when using either a Gaussian kernel or a uniform kernel. Data was taken from Chapter 7 in von Davier et al. (2004) and contained two 20-item test forms X and Y, where $N_X = 1,455$ and $N_Y = 1,453$ test takers took each form respectively under an EG design. Two different true equating transformations were generated, represented by G for Gaussian and U for Uniform, respectively. The reported values of $\hat{r}, \hat{s}$, given in Table 7.4 in von Davier et al. (2004) were used as true parameter values. Using the penalty function when selecting bandwidths, we had $h_x = 0.622$, and $h_y = 0.571$ for Gaussian, and $h_x = 1,003$, and $h_y = 1.003$ for uniform, which were used when obtaining the true equated scores.

From the true parameter values, we generated 1,000 score frequencies $N_J = (n_1, ..., n_J)$ and $N_K = (n_1, ..., n_K)$ from the multinomial distributions $Mult(N_J, r_1, ..., r_J)$ and $Mult(N_K, s_1, ..., s_K)$. The sample sizes used were identical to the sample sizes in the original data. In each replication, h_X and h_Y were estimated together with the r_j and s_k values. Although the first step is presmoothing, for simplicity the same log-linear model as was used for the true score probability parameters was considered. Thus, we used Eq. (3.4), with $T_r = 2$ for r_j and $T_s = 3$ for s_k. Instead of using the same model, an alternative would have been to either examine several competing log-linear models and use a fitting criterion (e.g., AIC, BIC, or likelihood ratio test) to choose the best model, as was done in, for example, Andersson and Wiberg (2017), or to skip the presmoothing step.

The estimated equated values were obtained using both the uniform and the Gaussian kernels. The estimated equated values were then compared with the true equated values, yielding a total of four scenarios. The words (uniform and gauss) indicate the kernel used in the estimations and the letters (G and U) indicate which kernel (Gaussian or uniform) was used in the true equating model. Figure 8.2 displays plots of the bias, MSE, RMSE, and SE. Note that there were only small differences between the four scenarios. The largest difference in bias was for the scenario when estimated equated values were obtained with the uniform kernel but the Gaussian kernel was used as the true model.

We can also calculate the simulated comparison indices for the four scenarios. The results are displayed in Table 8.3, where the smallest values on all indices were for uniform.U, closely followed by Gaussian.G. As these indices are relatively new, there are currently no guidelines on what is considered a small value, and this could therefore be a topic for future research.

Finally, APAD can be calculated for this example, as can be seen in Figure 8.3. The APAD had similar values for the four scenarios over almost the entire score range; the exception was for the highest score for the two scenarios when the estimated kernel and true kernel differed.

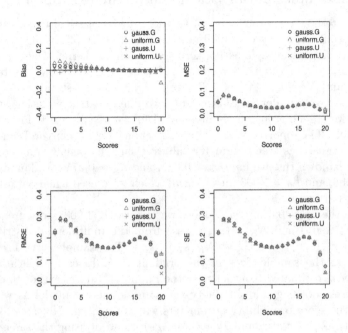

FIGURE 8.2: Bias, MSE, RMSE, and SE for the four scenarios, where the words (gauss and uniform) represent the kernel used in the estimated model and the suffix (U,G) represents the kernel in the true model.

8.4　Choice of Evaluation Measure

As seen in this chapter, there are many different evaluation measures. The DTM, PRE, and SEE look at different aspects of the equating transformation: practical significance, closeness to the discrete target distribution, and uncertainty in the equating transformation. Other measures are used to choose between two competing equating transformations based on the same data, such as the SEED, and cumulants are useful for comparing distributions. The equating indices are helpful tools for comparing different equating transformations. When performing simulation studies in which we can define a true equating transformation, the statistical measures bias, MSE, RMSE, and SE are good choices for comparing different equating methods. The choice of evaluation methods should thus be guided by the aim of a study, if empirical or simulated data have been used, and which aspects are most important to consider. We want to emphasize at this point the importance of using several measures, as they give different information.

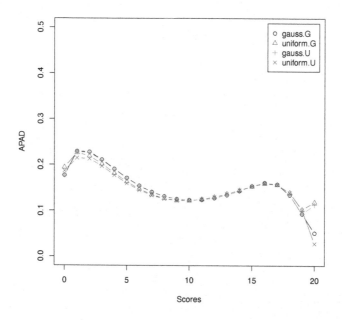

FIGURE 8.3: APAD for the four examined scenarios.

TABLE 8.3: Simulated statistical indices for the four scenarios.

Index	gauss.G	uniform.G	gauss.U	uniform.U
AMSD	0.0097	0.0135	0.0015	0.0053
AMAD	0.1582	0.1604	0.1610	0.1537
ARMSD	0.1811	0.1823	0.1829	0.1772

8.5 Summary

In this chapter, different evaluation criteria of the equating transformation have been discussed, including both equating-specific measures and statistical measures. A discussion on how to simulate test scores was also included. The choice of evaluation measures depends on the equating situation and, when evaluating an equating transformation, it is recommended to use several evaluation measures as they all give different information.

Some of the described evaluation measures were illustrated with simulated data. In the following two chapters, we will use real test data to illustrate and discuss some of the described evaluation measures of the equating transformation under both the EG and NEAT designs.

Part III

Applications

9

Examples under the EG design

The aim of this chapter is to illustrate some of the described methods for the different steps in the GKE framework when considering an EG design. To illustrate the choices, we use two test forms from a test administration of a private national evaluation system. The examples provided here have been selected to illustrate the flexibility of the GKE framework and are by no means exhaustive. We have also chosen to showcase choices for the GKE process in this chapter different from those that will be used in Chapter 10. The intention is to provide a practical illustration of the theory from the previous chapters along with explanations.

In the next section, we provide explanations for the software used in this chapter, followed by a brief description of the data and how to prepare these data for the selected software. We then proceed to the different GKE steps. In the first step, we illustrate how beta4 models and discrete kernel estimators can be employed for presmoothing the data. This is followed by a brief description on how to obtain the estimated score probabilities. The estimated score probabilities are then used in the continuization and equating steps to obtain bandwidth parameters and equating transformations, respectively, using both adaptive and Epanechnikov kernels.

Finally, we demonstrate how to evaluate the equating transformation using bootstrap SEE and the PRE measure. The use of Freeman-Tukey residuals is also included. The examples provided are selected for illustrative purposes, representing a subset of the possible choices. The overarching goal of this chapter is to guide the reader through these methods, providing **R** code and explanations of the obtained outputs.

9.1 Software Choice

In this chapter, **SNSequate** (González, 2014) is used to perform all the GKE steps. This choice was made as we wanted to include an illustration of the use of discrete kernel estimators as well as beta4 models in the presmoothing step, which is only possible with this package. Note that both **SNSequate** and **kequate** (Andersson et al., 2013) can handle log-linear models in the presmoothing step under the EG design.

As for the continuization step, both packages can handle uniform, logistic, and Gaussian kernels under the EG design. The reason for using **SNSequate** in this chapter is that we wanted to give an illustration of using the Epanechnikov and the adaptive kernels, described in Chapter 6, which are only implemented in this package. Note that both packages allow the selection of bandwidth by minimizing a penalty function in the EG design and also allow the use of Freeman-Tukey residuals. Both packages include measures, such as the PRE, for evaluating the equating transformation, as described in Chapter 8. **kequate** includes a built-in function call for using bootstrap SEE when IRT models have been used in the presmoothing step, which is illustrated in Chapter 10. There is no specific function for bootstrap SEE in **SNSequate**, so in this chapter we instead provide **R** code for calculating the bootstrap SEE. Note that the **R** code used in this chapter can be downloaded from https://github.com/MarieWiberg/GKE.

9.2   SEPA data

In this chapter, we will use data from a test developed by a private national evaluation system in Chile called SEPA (Sistema de Evaluación del Progreso en el Aprendizaje; System of Assessment Progress in Achievement), which is administered by the measurement center MIDE UC. SEPA consists of tests specifically designed to assess achievement in students from first to eleventh grades in the subjects of Language and Mathematics. The tests are given once a year. We used two test forms of the mathematics test, both composed of 50 items, and administered to $N_x = 1,458$ test takers (test form X) and $N_y = 2,619$ test takers (test form Y), respectively.

9.2.1   Preparing the SEPA Data for SNSequate

Under the EG design, the SEPA data correspond to two independent vectors of scores, one for each group taking the test forms X and Y, respectively (see Section 1.3.3). The SEPA data set is available in **SNSequate** and is a list containing two vectors with 1,458 test scores from form X (SEPA$xscores) and 2,619 test scores from form Y (SEPA$yscores). After the data have been loaded, score frequency distributions can be created using the freqtab() function from the **R** package **equate** (Albano, 2016) using the following code:

```
> data("SEPA", package = "SNSequate")
> library(equate)
> SEPAx <- freqtab(x = SEPA$xscores, scales = 0:50)
> SEPAy <- freqtab(x = SEPA$yscores, scales = 0:50)
```

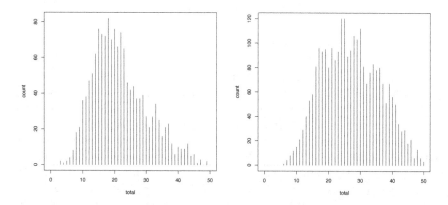

FIGURE 9.1: Frequency distribution of X (left) and Y (right) scores for the SEPA data under an EG design.

The arguments x and scales in freqtab() correspond in this case to a vector of scores and a single sequence of numbers indicating the score scale for the test, respectively. If an object of class freqtab is passed to the R function plot(), then with the equate package loaded, a plot of score frequencies will be displayed as vertical lines, as the ones shown in Figure 9.1. Basic descriptive statistics can also be obtained by passing the object to the summary() function, as it is shown in the following example code:

```
> rbind(x=summary(SEPAx),y=summary(SEPAy))
      mean       sd      skew      kurt min max    n
x 21.76200 8.509933 0.6379205 2.949851   3  49 1458
y 27.15426 8.947852 0.1730585 2.304413   6  50 2619
```

In this output, sd is the standard deviation, skew is the skewness, kurt is the kurtosis, min and max are the minimum and maximum test score values, and n is the number of test takers. According to these descriptive results, test form Y seems to be easier than test form X.

9.3 Step 1: Presmoothing

As seen in Chapter 3, we have several options for presmoothing the score data: log-linear models, nonparametric discrete kernel estimators, beta4 models, and IRT models. In this chapter, we concentrate on how to use beta4 models (Section 3.4.1) and the discrete kernel estimators (Section 3.3.1) for presmoothing, and postpone illustrations of the log-linear models and IRT models to Chapter 10.

9.3.1 Beta4 Models

The estimation of the score distribution using beta4 models is discussed in Lord (1965) and Hanson (1991a). These authors used the method of moments to estimate the four-parameter beta compound binomial model, considering the first four central moments of the score distribution (mean, standard deviation, skewness, and kurtosis). Just as with the moment-matching property of log-linear models which indicates that sample and fitted moments are matched, under the beta4 model the first four moments of the fitted distribution generally agree with those of the sample distribution. There are some exemptions in which fewer than four moments are matched, particularly when invalid parameter values are obtained. An example of an invalid parameter value occurs if the upper limit for proportion correct true score is above 1. For the cases when it is not possible to fit the four moments, Hanson (1991a) proposed a method that fits the first three moments, making the fourth moment of the fitted distribution as close as possible to the observed one.

To perform presmoothing with a beta4 model, as described in Section 3.4.1, we use the `BB.smooth()` function in **SNSequate**. A typical call to this function reads as

```
> BB.smooth(x, nparm, rel)
```

where x is a score frequency distribution, which can be created with the `freqtab()` function from the **equate** package, and nparm indicates the number of parameters of the beta distribution to be estimated (2 or 4). As described in Section 3.4.1, it can be assumed that the conditional distribution of observed scores given the true score is either binomial or compound binomial. The `rel` argument is used to specify the conditional distribution; when `rel=0`, the binomial distribution is used; otherwise the compound binomial distribution is assumed.

As an example, the following code can be used to estimate the beta4 model when the conditional distribution of scores given the true score is binomial:

```
> beta4nx <- BB.smooth(SEPAx,nparm=4,rel=0)
> beta4ny <- BB.smooth(SEPAy,nparm=4,rel=0)
```

We use the `beta4nx` object to illustrate how the output looks, although we omitted rows 4 to 49 here.

```
> beta4nx
```

```
Call:
BB.smooth.default(x = SEPAx, nparm = 4, rel = 0)
```

Estimated frequencies and score probabilities:

```
      Est.Freq.  Est.Score.Prob.
1   1.469639e-04    1.007983e-07
2   2.405856e-03    1.650107e-06
3   1.940196e-02    1.330725e-05
-   --------       -----------
50  7.495353e-01    5.140846e-04
51  2.775240e-01    1.903457e-04
```

Parameters:

```
[1] 0.9692559 2.6985800 0.2323938 1.0000000
```

The output shows estimated frequencies and estimated score probabilities in the columns Est.Freq. and Est.Score.Prob., respectively. These values are stored in the objects freq.est and prob.est, and can be accessed using the dollar sign (see example below). The output also gives the parameter values for the beta4 model, where the first and second values are the shape parameter values of the beta distribution (usually labelled α and β in standard statistics textbooks) and the third and fourth values are the lower and upper limits. In this case, only the first three parameters could be fit, and thus the upper limit was set to 1, as explained above. For a general discussion of different beta4 models, refer to Chapter 9 in Brennan et al. (2009).

9.3.2 Discrete Kernel Estimators

Discrete kernel estimators that can be used for presmoothing were described in Section 3.3.1. As mentioned there, for count data, the discrete triangular has been suggested to be used with large sample sizes, while the binomial has been suggested to be used with smaller sample sizes. The Dirac uniform (DiracDU) kernels have also been suggested to be used with categorical data. Nevertheless, we examine all these discrete kernel estimators on the same data here.

To presmooth score distributions with these three different kernels, we use the discrete.smooth() function implemented in the **SNSequate** package. A typical call to the function reads as

```
> discrete.smooth(scores, kert, h,...)
```

where the argument scores is used to feed the score data. Further, the kert argument states which type of discrete kernel is to be used, with the options "dirDU" for DiracDU, "bino" for Binomial, and "triang" for discrete triangular, respectively. Finally, h is used to provide a value for the bandwidth parameter.

The following code can be used to fit the binomial and triangular discrete kernels when $h = 0.25$, as well as the naive kernel obtained as a particular case of the Dirac discrete uniform kernel when $h = 0$ (see Section 3.3.1), for score data coming from test form X.

```
> psxB <- discrete.smooth(scores=rep(0:50,SEPAx),kert="bino",
+ h=0.25,x=0:50)
> psxD <- discrete.smooth(scores=rep(0:50,SEPAx),kert="dirDU",
+ h=0.0,x=0:50)
> psxT <- discrete.smooth(scores=rep(0:50,SEPAx),kert="triang",
+ h=0.25,x=0:50)
```

Similar objects can be created for test form Y by exchanging the data with SEPAy. Each call to the `discrete.smooth()` function includes specifying the vector of scores (`scores`), the chosen kernel (`kert`), and the size of the bandwidth (`h`). The argument x is used to specify the points at which the probability mass function (PMF) is to be estimated, in this case, the whole score scale range.

An example output looks as follows:

```
> psxB

Call:
discrete.smooth.default(scores = rep(0:50, SEPAx),kert = "bino",
    h = 0.25, x = 0:50)

Estimated score probabilities:

    Est.Score.Prob.
1       0.0000000000
2       0.0000000000
3       0.0005788037
4       0.0008508820
-       ----------
49      0.0009687621
50      0.0005653135
51      0.0001960208
```

The column `Est.Score.Prob.` shows the estimated score probabilities under the chosen discrete kernel (the binomial in this case). The estimated score probabilities between row 5 and 48 have been omitted.

9.4 Step 2: Estimating Score Probabilities

Because under the EG design the two independent score vectors come from univariate score distributions, there is no need to map the estimated probabilities into *marginal* estimated score probabilities. As seen in Section 4.1.1, the DF for the EG design is just an identity mapping, which means that the estimated score probabilities can be obtained directly from the beta4 and discrete kernels outputs, respectively. We illustrate how to obtain the estimated score probabilities in the next two subsections.

9.4.1 Beta4 Models

For the beta4 model, the estimated score probabilities r_j and s_k are stored in the prob.est argument of the obtained beta4 objects. The following code can be used with the previously created objects beta4nx and beta4ny to extract the r_j and s_k estimated score probabilities. The outputs for test scores 5 to 44 are omitted below.

```
> rj.b4 <- beta4nx$prob.est
> sk.b4 <- beta4ny$prob.est
> cbind(0:50, rj.b4, sk.b4)
           rj.b4          sk.b4
 [1,]  0 1.007983e-07 1.654318e-08
 [2,]  1 1.650107e-06 2.786345e-07
 [3,]  2 1.330725e-05 2.322996e-06
 [4,]  3 7.055688e-05 1.279238e-05
 [5,]  4 2.769029e-04 5.240505e-05
 ---   -- ------------ ------------
[46,] 45 2.851895e-03 8.834713e-03
[47,] 46 2.126776e-03 6.904805e-03
[48,] 47 1.491515e-03 5.076551e-03
[49,] 48 9.514931e-04 3.389156e-03
[50,] 49 5.140846e-04 1.899119e-03
[51,] 50 1.903457e-04 7.043412e-04
```

A way to evaluate the quality of the presmoothing step is using graphics. This can be done by plotting the presmoothed and the observed frequency distributions together. Figure 9.2 illustrates presmoothing of the right-skewed X frequency distribution and the more symmetric Y frequency distribution when using beta4 models, and it was produced using the code shown below. From this figure, it can be seen that the beta4 model fit captures the main features of the observed frequency distributions.

```
> plot(0:50,as.matrix(SEPAx)/sum(as.matrix(SEPAx)),type="h",
```

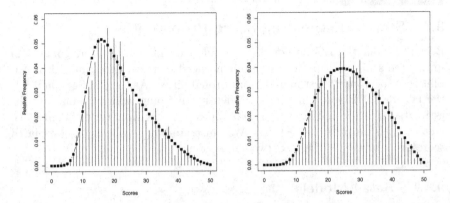

FIGURE 9.2: Presmoothing of frequency distribution of X scores (left) and Y scores (right) scores for the `SEPA` data using beta4 models.

```
+ ylim=c(0,0.06),ylab="Relative Frequency",xlab="Scores")
> lines(0:50,rj.b4,type="b",pch=15)

> plot(0:50,as.matrix(SEPAy)/sum(as.matrix(SEPAy)),type="h",
+ ylim=c(0,0.06),ylab="Relative Frequency",xlab="Scores")
> lines(0:50,sk.b4,type="b",pch=15)
```

9.4.2 Discrete Kernel Estimators

For discrete kernel presmoothing under the EG design, the estimated score probabilities r_j and s_k can be obtained directly using the `prob.est` argument of the object created using the `discrete.smooth()` function. Using the previously created object `psxB` and a similarly created object for test form Y labelled `psyB`, the estimated score probabilities for a binomial discrete kernel presmoothing are obtained using the following code. Note that we have chosen to omit the display of scores 6 to 44 in the output.

```
> rj.dkb <- psxB$prob.est
> sk.dkb <- psyB$prob.est
> cbind(0:50,rj.dkb,sk.dkb)
        rj.dkb       sk.dkb
[1,]  0 0.0000000000 0.0000000000
[2,]  1 0.0000000000 0.0000000000
[3,]  2 0.0005788037 0.0000000000
[4,]  3 0.0008508820 0.0000000000
[5,]  4 0.0010669000 0.0000000000
[6,]  5 0.0018893995 0.0001714256
```

```
---      --  ------------ ------------
[46,] 45 0.0025615265 0.0081448322
[47,] 46 0.0019257205 0.0052981215
[48,] 47 0.0009887183 0.0045197855
[49,] 48 0.0009687621 0.0034690647
[50,] 49 0.0005653135 0.0021483227
[51,] 50 0.0001960208 0.0008976741
```

Again, we are interested in visualizing the quality of the presmoothing by plotting the observed frequency distributions together with the fitted ones. The following code is used to illustrate the presmoothed distributions of X and Y in comparison with the observed score frequency distributions, as is shown in Figure 9.3. Note that in the below code some objects (rj.dkd, rj.dkt, sk.dkd, and sk.dkt) had not been used earlier in this chapter, but they can be obtained or created in a similar way as was done for rj.dkb. For instance, using the previously created psxD and psxT objects, rj.dkd and rj.dkt can be extracted using the dollar sign. On the other hand, using the frequency distribution for Y (SEPAy), both the psyD and psyT can first be created to then obtain sk.dkd and sk.dkt using the dollar sign.

```
> plot(0:50,as.matrix(SEPAx)/sum(as.matrix(SEPAx)),lwd=2.0,
+ xlab="Scores",ylab="Relative Frequency",type="h")
> points(0:50,rj.dkb,type="b",pch=0)
> points(0:50,rj.dkd,type="b",pch=1)
> points(0:50,rj.dkt,type="b",pch=2)
> legend("topright",pch=c(16,0,1,2),lty=c(1,1,1,1),
+ c("Observed","Binomial","Dirac","Triangular"))

> plot(0:50,as.matrix(SEPAy)/sum(as.matrix(SEPAy)),lwd=2.0,
+ xlab="Scores",ylab="Relative Frequency",type="h")
> points(0:50,sk.dkb,type="b",pch=0)
> points(0:50,sk.dkd,type="b",pch=1)
> points(0:50,sk.dkt,type="b",pch=2)
> legend("topright",pch=c(16,0,1,2),lty=c(1,1,1,1),
+ c("Observed","Binomial","Dirac","Triangular"))
```

The presmoothing of the frequency distribution of X and Y using the three different kernels as compared with the observed score frequencies is shown in Figure 9.3. Note that all three discrete kernels gave similar results in our example, and they all produce score distributions that closely resemble the observed score distributions. Also note that these presmoothed distributions are quite different in comparison to those obtained using beta4 models for presmoothing, as shown in Figure 9.2, where only the main features of the data were captured, and not all irregularities. It is possible, although we have not examined this, that the discrete kernel presmoothing curves could have been more similar to the beta4 presmoothing curves if another value for the bandwidth parameter would have been used.

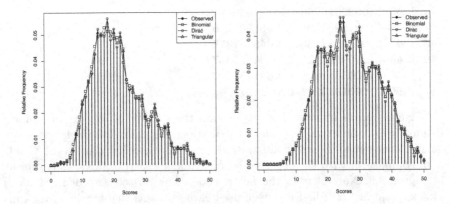

FIGURE 9.3: Presmoothing of frequency distribution of X scores (left) and Y scores (right) for the SEPA data using three discrete kernels.

9.5   Step 3: Continuization

The continuization step, generally described in Chapter 5, involves the use of both a continuous random variable (characterizing the kernel to be used as described in Section 5.4) and a bandwidth parameter that controls the degree of smoothness in the continuization (see Chapter 6). In this chapter, we will use the penalty function (described in Section 6.3) for the bandwidth selection and the Epanechnikov kernel (described in Section 5.4.4) and the adaptive kernels (described in Section 5.4.5). We will use the functions `bandwidth()` and `ker.eq()` from **SNSequate**.

9.5.1   Bandwidth Selection

The `bandwidth()` function implements the penalty method described in Section 6.3 to obtain the bandwidth parameter h_X. It has the following typical call:

```
bandwidth(scores, kert, design, r, ...)
```

where the argument `scores` is used to specify the score data as raw sample frequencies. The argument `kert` refers to the kernel type (with options `gauss`, `logis`, `unif`, `epan`, and `adap` for the Gaussian, logistic, uniform, Epanechnikov, and adaptive kernels, all described in Section 5.4). Further, the argument `design` is used to specify the data collection design (with options `EG`, `SG`, `CB`, `EG`, `NEATCE`, and `NEATPSE`, for the EG, SG, CB, and NEAT chained equating and NEAT poststratification equating). The argument `r` is used to provide the estimated score probabilities needed in the penalty function to be minimized in Eq. (6.7).

The bandwidth() function also incorporates the degree argument when presmoothing using log-linear models, which is the default option. Because the ker.eq() function that will be described next makes internal calls to the bandwidth() function and because they share some arguments that will be described in Section 9.6, a complete description is suppressed here.

The following code illustrates how to obtain the bandwidth values using the penalty method, when different inputs are considered for both the estimated score probabilities and the densities involved in the penalty function to be minimized. For the estimated score probabilities we use those obtained using the binomial discrete kernel presmoothing (see Section 9.4.2), the beta4 (see Section 9.4.1), and the sample score probabilities obtained from the observed data. We consider the densities derived when using both the Epanechnikov and the adaptive kernels. In this code, we use the argument r to provide estimated score probabilities obtained under the different methods mentioned above. Note that the observed score probabilities rj.obs and sk.obs are obtained directly from the observed data and can be computed using the code in the first two rows below. Results are summarized in Table 9.1.

```
> rj.obs <- as.matrix(SEPAx)/sum(as.matrix(SEPAx))
> sk.obs <- as.matrix(SEPAy)/sum(as.matrix(SEPAy))

> hx.obs.ep <- bandwidth(scores=as.vector(SEPAx),kert="epan",
+ design="EG",r=rj.obs)$h
> hx.b4.ep <- bandwidth(scores=as.vector(SEPAx),kert="epan",
+ design="EG",r=rj.b4)$h
> hx.dkb.ep <- bandwidth(scores=as.vector(SEPAx),kert="epan",
+ design="EG",r=rj.dkb)$h

> hx.obs.ad <- bandwidth(scores=as.vector(SEPAx),kert="adap",
+ design="EG",r=rj.obs)$h
> hx.b4.ad <- bandwidth(scores=as.vector(SEPAx),kert="adap",
+ design="EG",r=rj.b4)$h
> hx.dkb.ad <- bandwidth(scores=as.vector(SEPAx),kert="adap",
+ design="EG",r=rj.dkb)$h

> hy.obs.ep <- bandwidth(scores=as.vector(SEPAy),kert="epan",
+ design="EG",r=sk.obs)$h
> hy.b4.ep <- bandwidth(scores=as.vector(SEPAy),kert="epan",
+ design="EG",r=sk.b4)$h
> hy.dkb.ep <- bandwidth(scores=as.vector(SEPAy),kert="epan",
+ design="EG",r=sk.dkb)$h

> hy.obs.ad <- bandwidth(scores=as.vector(SEPAy),kert="adap",
+ design="EG",r=sk.obs)$h
```

```
> hy.b4.ad <- bandwidth(scores=as.vector(SEPAy),kert="adap",
+ design="EG",r=sk.b4)$h
> hy.dkb.ad <- bandwidth(scores=as.vector(SEPAy),kert="adap",
+ design="EG",r=sk.dkb)$h
```

TABLE 9.1: Bandwidth selected for Epanechnikov and adaptive kernels using the penalty method for score probabilities obtained from observed samples, the binomial discrete kernel estimator, and a beta4 model.

	h_X			h_Y		
	Observed	Binomial	Beta4	Observed	Binomial	Beta4
Epanechnikov	2.46	2.71	2.75	2.31	2.25	2.75
Adaptive	1.28	1.14	0.59	1.63	1.40	0.64

From Table 9.1, it can be seen that bandwidths obtained using the observed sample probabilities and those obtained using the binomial discrete kernel are larger than the ones obtained when using beta4 score probabilities in both X and Y scores for adaptive kernels. For the Epanechnikov kernel, the bandwidth values are quite homogeneous for both X and Y scores. In the next section, we will evaluate how these values impact the actual equating.

9.5.2 Kernel Selection

In Section 5.4, five different kernels were described that can be used within GKE: Gaussian, logistic, uniform, Epanechnikov, and adaptive kernels. Although **SNSequate** can handle all these different kernels, we have chosen here to illustrate the use of the Epanechnikov kernel (see Section 5.4.4) and adaptive kernels (see Section 5.4.5) for continuization. The choice of kernel as well as a chosen bandwidth can be directly fed into the KE function ker.eq(), which is described in the following section.

9.6   Step 4: Equating

The equating step was described in Chapter 7 and is performed in **SNSequate** using the ker.eq() function. A typical call to ker.eq() has the following general structure

```
ker.eq(scores, kert, hx = NULL, hy = NULL,
alpha=NULL, h.adap=NULL, r, s, ...)
```

Both the scores argument and the kert argument for choosing kernels have the same options as described for the bandwidth() function in Section

9.5.1. The bandwidth parameters can be set using the arguments hx and hy. If not provided (default), these values are automatically calculated using the penalty method. Either the r and s arguments must be filled in, or otherwise the degree argument must be filled in, so that score probabilities are calculated internally using polynomial log-linear models.

In order to use adaptive kernel continuization, the sensitivity parameter α and a bandwidth h_X for the pilot estimate are needed, as described in Section 5.4.5. The sensitivity parameter α is specified using the argument alpha, and the value of the bandwidth parameter h_X for the pilot estimate is set using the argument h.adap. If no value is provided for these two arguments, then the default value for alpha is set to 0.5 and h_X will be estimated using the penalty method. A complete description of all the currently implemented arguments in ker.eq() is shown in Table 9.2

The following code is used to perform equating under the EG design using the Epanechnikov kernel continuization with the score probabilities obtained from the beta4 model (Section 9.4.1) and the binomial discrete kernel presmoothing (Section 9.4.2). Note that the first line of code combines test X and test Y data into a matrix before the equating is carried out. The subsequent lines perform the equating. The last lines compare the equated scores obtained from the two equating transformations, yielding the table output of the equated test scores rounded to four decimal places (the results for scores 5 to 44 are omitted).

```
> SEPAmat <- cbind(SEPAx,SEPAy)
> eq.epan.b4 <- ker.eq(scores=SEPAmat,hx=hx.b4.ep,hy=hy.b4.ep,
+ kert="epan", design="EG",r=rj.b4 ,s=sk.b4)
> eq.epan.dkb <- ker.eq(scores=SEPAmat,hx=hx.dkb.ep,hy=hy.dkb.ep,
+ kert="epan", design="EG",r=rj.dkb,s=sk.dkb)

> comp.epan <- cbind(Scale=0:50,Epan.b4=eq.epan.b4$eqYx,
+ Epan.dkb=eq.epan.dkb$eqYx)
> round(comp.epan,4)
      Scale Epan.b4 Epan.dkb
 [1,]     0  0.8969   3.6737
 [2,]     1  2.0359   4.6873
 [3,]     2  3.1907   5.4870
 [4,]     3  4.3634   6.2033
 [5,]     4  5.5518   6.8987
 ----     -  ------   ------
[46,]    45 47.4960  47.6465
[47,]    46 48.1751  48.3747
[48,]    47 48.8656  48.9987
[49,]    48 49.5565  49.6091
[50,]    49 50.2561  50.2127
[51,]    50 50.9813  50.8309
```

TABLE 9.2: List of arguments in `ker.eq()`.

Argument(s)	Designs	Description
scores	ALL	If "EG" design is specified, a vector containing the raw sample frequencies coming from one group taking the test.
kert	ALL	The type of kernel used for continuization. Current options include "gauss", "logis", "uniform", "epan", and "adap" for the Gaussian, logistic, uniform, Epanechnikov, and adaptive kernels, respectively.
hx,hy	ALL	Value of the bandwidth parameter used for continuization of $F(x)/F(y)$. If not given (default), it is automatically calculated.
degree	EG	For the EG design, this indicates the number of power moments to be fitted to the marginal distributions when using log-linear models.
design	ALL	The equating design. Options are EG, SG, CB, NEAT_CE, and NEAT_PSE.
Kp	ALL	Weight for the second term in the combined penalization function used to obtain h.
scores2	CB, NEAT CE, NEAT PSE	A vector of sample frequencies.
degreeXA	NEAT CE, NEAT PSE	Number of interactive power moments used with test forms X and A.
degreeYA	NEAT CE, NEAT PSE	Number of interactive power moments used with test forms Y and A.
J	CB, NEAT CE, NEAT PSE	The number of possible X scores.
K	CB, NEAT CE, NEAT PSE	The number of possible Y scores.
L	CB, NEAT CE, NEAT PSE	The number of possible A scores.
wx,wy	CB	Weights.
w	NEAT	Weight for the synthetic population.
gapsX,gapsY,gapsA	NEAT	A list containing: index, a vector of indices; degree, an integer indicating the maximum degree of the moments fitted.
lumpX,lumpY,lumpA	NEAT	An integer to represent the index where an artificial "lump" is created in the marginal distribution of frequencies.
alpha	EG	Sensitivity parameter. Only for adaptive kernel.
h.adap	EG	A list with bandwidths (hx,hy) for adaptive kernel for each test form.

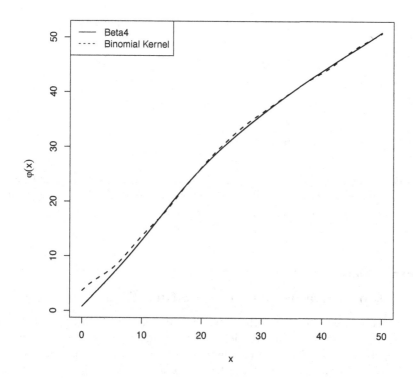

FIGURE 9.4: Equating transformations using the Epanechnikov kernel continuization and estimated score probabilities weights obtained with both the beta4 and binomial discrete kernel presmoothing.

A graphical comparison of the equating transformations is given in Figure 9.4. In this figure, we can see that the equating transformations are similar over most of the scores scale but differ somewhat from each other in the lower score range.

Next, we compare the equating transformations using the Epanechnikov and the adaptive kernel kernel continuization, and with score probability weights obtained using beta4 presmoothing (Section 9.4.1). For the former case, we used the previously obtained object **eq.epan.b4**, and for the latter we created the object **eq.adap.b4** in the first two lines of code. The subsequent three lines of code are used to display the numerical comparison.

```
> eq.adap.b4 <- ker.eq(scores=SEPAmat,hx=hx.b4.ad,hy=hy.b4.ad,
+ kert="adap",design="EG",r=rj.b4,s=sk.b4,alpha=0.1)

> comp.epan.adap <- cbind(Scale=0:50,Epan.b4=eq.epan.b4$eqYx,
```

```
+ Adap.b4=eq.adap.b4$eqYx)
> round(comp.epan.adap,4)
```

The output showing the comparison of the first five equated scores is given below.

```
     Scale Epan.b4 Adap.b4
[1,]     0  0.8969  0.7460
[2,]     1  2.0359  1.8849
[3,]     2  3.1907  3.0312
[4,]     3  4.3634  4.1879
[5,]     4  5.5518  5.3590
---
```

Overall the equated values are quite similar. A graphical representation of this comparison is shown in Figure 9.5, where it is evident that the equating transformations for these cases are very similar.

9.7 Step 5: Evaluating the Equating Transformation

As we have seen in Chapter 8, there are several measures that can be used to evaluate the equating transformation and the quality of the equated values. In this section, we have chosen to illustrate two evaluation measures that are useful when evaluating an equating transformation. We start by illustrating how to obtain the percent relative error (PRE). Next, we illustrate how to calculate bootstrap standard error of equating (SEE) for the equating conducted previously using the Epanechnikov kernel. We have also included an illustration of how to compute Freeman-Tukey residuals (described in Section 3.5.1) to assess model fit in the presmoothing step. The reason for placing this example in this section is that we need the objects created in Section 9.6 to feed the gof() function that will be used to obtain Freeman-Tukey residuals.

9.7.1 PRE

One way of comparing different equating transformations is through the moments of the score distributions. As seen in Section 8.1.2, the PRE measures the difference between the moments of the distribution for equated values with those of the score distribution to which scores are being equated. The function PREp() receives as arguments an equating object and the number of moments to be evaluated. For example, the following code is used to obtain a comparison of the distributions when equating using the Epanechnikov kernel with both the beta4 and the binomial discrete kernel presmoothing, and the adaptive kernel with the beta4 presmoothing for the first ten moments, i.e., $B = 10$ in Eq. (8.1).

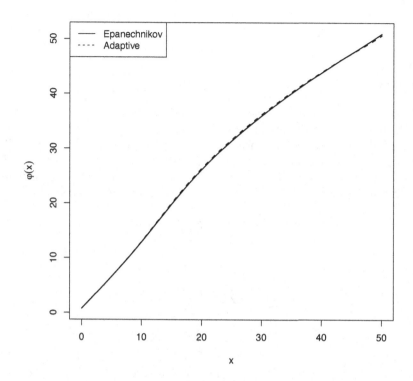

FIGURE 9.5: Equating transformations using the Epanechnikov and the adaptive kernel continuization with the beta4 presmoothing.

```
> data.frame(Epan.b4=round(PREp(eq.epan.b4,10)$preYx,4),
+ Epan.dkb=round(PREp(eq.epan.dkb,10)$preYx,4),
+ Adap.b4=round(PREp(eq.adap.b4,10)$preYx,4))

   Epan.b4 Epan.dkb Adap.b4
1  -0.0080  -0.0188  0.0000
2  -0.0447  -0.0589 -0.0058
3  -0.0436  -0.0636 -0.0050
4   0.0107  -0.0168  0.0064
5   0.1147   0.0789  0.0288
6   0.2619   0.2147  0.0618
7   0.4468   0.3815  0.1053
8   0.6656   0.5718  0.1591
9   0.9165   0.7800  0.2233
10  1.1982   1.0024  0.2982
```

The adaptive kernel with the beta4 presmoothing model (`Adap.b4`) had the smallest PRE in all 10 moments, followed by using the discrete binomial kernel in the presmoothing with the Epanechnikov kernel (`Epan.dkb`). Using a beta4 model in the presmoothing and an Epanechnikov kernel (`Epan.b4`) had the highest PRE values in all 10 moments.

9.7.2   Bootstrap Standard Error of Equating

Another way of comparing different equating transformations is to compare their SEE. The algorithm used to obtain bootstrap SEE was described in Section 8.1.5 for an equating transformation $\varphi(x)$. The bootstrap SEE is typically used when we do not have analytical expressions for the SEE. In the GKE framework, we have shown in Chapter 7 that $\varphi(x)$ can be re-written more explicitly in terms of bandwidths and estimated scores probabilities as

$$\varphi(x; \hat{\boldsymbol{r}}, \hat{\boldsymbol{s}}) = F_{h_Y}^{-1}(F_{h_X}(x; \hat{\boldsymbol{r}}); \hat{\boldsymbol{s}}),$$

where

$$F_{h_X}(x) = \sum_j \hat{r}_j K(R_{jX}(x)),$$

$$F_{h_Y}(y) = \sum_k \hat{s}_k K(R_{kY}(y)).$$

As the values of h_X and h_Y depend on those of r_j and s_k (see Eq. (6.5)), each time a new bootstrap sample is drawn, new values for these parameters are used to build the equating transformation.

The code below can be used to implement the bootstrap algorithm to obtain bootstrap SEE. First, the score data is read in and the number of test takers who took test forms X and Y, respectively are stored in `Nx` and `Ny`, respectively. Next, we set the number of replicates to $L = 100$, followed by allocating memory for the equating transformation to `matboot`. Within the bootstrap algorithm, the data is first sampled with replacement using `sample` and then score data is placed into frequency tables using `freqtab` from the R package **equate**. Next, the beta4 model is used to obtain estimated score probabilities for test form X and test form Y, followed by using these score probabilities and the Epanechnikov kernel when selecting the bandwidths. This is followed by performing the equating and saving it in `matboot`. Finally, the bootstrap SEEs are calculated in the final row by calculating the standard deviation of the equated values.

```
> scores.x <- rep(0:50,SEPAx)
> scores.y <- rep(0:50,SEPAy)

> Nx <- length(scores.x)
```

```
> Ny <- length(scores.y)
> L  <- 100

> matboot <- matrix(NA,ncol=51,nrow=L)

> set.seed(234)
> i  <- 0
> it <- 0
> while(it<L){
> i  <-i+1
> x.b <-sample(scores.x,size=Nx,replace=TRUE)
> y.b <-sample(scores.y,size=Ny,replace=TRUE)
> SEPAx.b  <-freqtab(x.b,scale=0:50)
> SEPAy.b  <-freqtab(y.b,scale=0:50)
> SEPAmat.b<-cbind(SEPAx.b,SEPAy.b)

> beta4nx.b<- BB.smooth(SEPAx.b,nparm=4,rel=0)
> beta4ny.b<- BB.smooth(SEPAy.b,nparm=4,rel=0)
> rj.b4.b  <- beta4nx.b$prob.est
> sk.b4.b  <- beta4ny.b$prob.est

> hx.b4.ep.b <- bandwidth(scores=as.vector(SEPAx.b),
+ kert="epan", design="EG",r=rj.b4.b)$h
> hy.b4.ep.b <- bandwidth(scores=as.vector(SEPAy.b),
+ kert="epan", design="EG",r=sk.b4.b)$h

> eq.epan.b4 <- tryCatch(ker.eq(scores=SEPAmat.b,hx=hx.b4.ep.b,
+ hy=hy.b4.ep.b, kert="epan",design="EG",r=rj.b4.b,
+ s=sk.b4.b),error=function(e)NULL)
> if(!is.null(eq.epan.b4)){
> it=it+1
> matboot[it,]<-eq.epan.b4$eqYx
> }
> print(i)
> print(it)
> }
> apply(matboot,2,sd)
```

A graphical representation of the bootstrap SEE is shown in Figure 9.6. In this figure, we can see that the bootstrap SEEs are largest around score value 40 and smaller for lower scores and for very high scores.

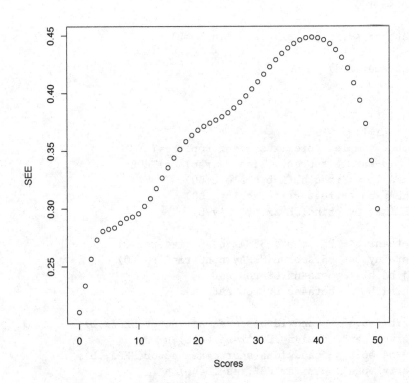

FIGURE 9.6: Bootstrap SEE when using the beta4 model for presmoothing and the Epanechnikov kernel continuization.

9.7.3 Freeman-Tukey Residuals

In Figures 9.2 and 9.3, fitted score probabilities were compared with observed (sample) score probabilities, as a way to examine model fit. A related alternative approach compares predicted frequencies with the observed frequencies. Freeman-Tukey residuals, as described in Section 3.5.1, are a useful tool for assessing differences between observed and predicted frequencies. The gof() function in **SNSequate** implements three goodness-of-fit measures, with options FT, Chisq, and KL for Freeman-Tukey residuals, the Pearson chi-squared test, and the symmetrised Kullback-Leibler divergence.

The following code can be used to obtain the Freeman-Tukey residuals from the three equating objects previously created.

```
> FTr.epan.b4 <-gof(SEPAx,eq.epan.b4$rj*eq.epan.b4$nx,"FT")
> FTr.epan.dkb<-gof(SEPAx,eq.epan.dkb$rj*eq.epan.dkb$nx,"FT")
> FTr.adap.b4 <-gof(SEPAx,eq.adap.b4$rj*eq.adap.b4$nx,"FT")
```

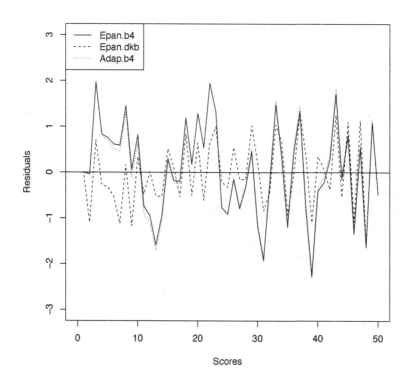

FIGURE 9.7: Freeman-Tukey residuals.

In the code above, the gof() arguments are the observed frequencies (SEPAx), the fitted frequencies (obtained by multiplying the estimated probabilities with the number of test takers), and the method choice for the goodness-of-fit, which in this case to obtain Freeman-Tukey residuals is FT. The following code is used to obtain Figure 9.7, which is a plot of the Freeman-Tukey residuals.

```
> plot(0:50,FTr.epan.b4$ft.res,type="l",lty=1,
+ ylab="Residuals",xlab="Scores",ylim=c(-3,3))
> lines(0:50,FTr.epan.dkb$ft.res,type="l",lty=2)
> lines(0:50,FTr.adap.b4$ft.res,type="l",lty=3)
> legend("topleft",lty=c(1,2,3),c("Epan.b4","Epan.dkb","Adap.b4"))
> abline(h=0)
```

From Figure 9.7, it can be seen that in all cases, the residuals show no particular pattern and lie between ±3, as expected. Also note that using either Epanechnikov or adaptive kernels with beta4 presmoothing yielded similar Freeman-Tukey residuals, whereas using an Epanechnikov kernel with binomial discrete kernel presmoothing yielded slightly different Freeman-Tukey residuals.

9.8   Summary

In this chapter, we have given examples of how to use GKE with different alternatives at each step of the GKE process under an EG design. Throughout the chapter, we used samples from two test forms which produced binary-scored data. The **R** package **SNSequate** was used for the different GKE steps. Note that several of the illustrated options in this chapter are also available if a NEAT design is considered instead. Even though many options were illustrated in this chapter, there are others available, including different combinations at each step in the GKE framework, that could have been used instead. Some of these other options in GKE are illustrated in the next chapter when conducting equating under the NEAT design.

10

Examples under the NEAT design

In this chapter, the aim is to demonstrate some of the various options available at each step in the GKE framework, under the NEAT design, and for both binary and polytomously scored items. To illustrate the choices for binary-scored items, two samples from a real college admissions test are utilized. In the next sections, a brief description is given of the software choice as well as of the admission test and its purpose. For polytomous item responses, simulated data is used, which is introduced in the third section of the chapter. The illustrations showcase the application of GKE in different scenarios, emphasizing its flexibility and adaptability based on the nature of the test items.

Certainly, the options presented in this chapter for both binary and polytomously scored items are not exhaustive. The intention is to provide a practical illustration of some of the useful choices that have been theoretically described in previous chapters. By going through the procedures and explaining the outcomes, the chapter aims to offer a comprehensive understanding of the application of the GKE framework under the NEAT design.

Two methods for presmoothing score distributions will be illustrated: the use of log-linear models and IRT models. Additionally, different kernels (Gaussian, uniform and logistic) will be explored, and examples will be provided for different bandwidth selection methods, such as cross-validation and double smoothing. Finally, the equating transformation will be evaluated using various measures, including PRE, SEE, SEED, bootstrap SEE, MSD, MAD, and RMSD. These evaluations will provide a comprehensive understanding of the performance of the equating transformation when different methods have been used.

10.1 Software Choice

In this chapter, **kequate** (Andersson et al., 2013) is used to perform all the GKE steps. This choice was made as we wanted to include an illustration of when we use both binary and polytomous IRT models in the presmoothing step, which is only possible with this package. Note that both **kequate** and **SNSequate** can handle log-linear models in the presmoothing step under the NEAT design, although they differ in their implementations. When using

kequate the standard `glm()` function in **R** can be used to build sophisticated log-linear models that can then easily be read into the package, while in **SNSequate** there is a built-in function for fitting log-linear models.

As for the continuization step, both packages can handle uniform, logistic, and Gaussian kernels under the NEAT design. However, another reason for using **kequate** in this chapter is that we wanted to illustrate some of the described bandwidth selection methods in Chapter 6 that are currently implemented only in **kequate**. Although both packages implement the penalty method when selecting bandwidths, **kequate** also offers the choice of using double smoothing and cross-validation. Also, both packages offer common evaluation measures within the NEAT design, such as SEE, SEED, and PRE, as described in Chapter 8.

To plot the marginal frequencies of the unique and anchor test forms we used the **R** package **equate** (Albano, 2016), and to construct a simulated data example, we used the **R** package **ltm** (Rizopoulos, 2006). For fitting both binary and polytomous IRT models we used the **R** package **mirt** (Chalmers, 2012) and for examining IRT model fit we used the **R** package **psych** (Revelle, 2023). Note that the **R** code used in this chapter can be downloaded from https://github.com/MarieWiberg/GKE.

10.2 ADM Data

In this chapter, the analyses will be conducted using samples from two test forms of a college admissions test, denoted as the ADM data. The purpose of this test is to assess the knowledge and skills acquired during compulsory school and upper-secondary school, which are essential for higher education. The test follows a norm-referenced format, ranking test takers based on their test results. It is administered twice a year, and individuals have the option to retake the test multiple times, with only the best results considered in the selection process for higher education.

The college admissions test comprises separate verbal and quantitative sections, each consisting of 80 multiple-choice items, all scored in binary format. These sections are equated separately. Within each section, the 80 items are divided into two parts, each constructed based on the same test specifications. To facilitate equating for the verbal section, a 40-item external verbal anchor test is administered to a smaller subset of test takers during each test administration. Similarly, a 40-item external quantitative anchor test is given to a smaller subset of test takers to aid in the equating process for the quantitative section.

The data sets pertaining to these two test forms, as well as an external anchor test form, are available in the ADM file folder, accessible at the following GitHub repository: https://github.com/MarieWiberg/GKE. Note that these data files should be placed in the current **R** working directory being used to run the code given in this chapter.

The ADMX data contain a sample of 2,000 test takers from population $\mathcal{P}$ who took a verbal test form X (with 80 items) and the external verbal anchor test form A (with 40 items), whereas the ADMY data contain a sample of 2,000 test takers from population $\mathcal{Q}$ who took a verbal test form Y (with 80 items) and the same external anchor test form A. In summary, each data set comprises matrices of size 2000 × 120, where the first 40 columns contain the binary-scored answers to the external verbal anchor items and the remaining 80 columns are the binary-scored answers to the verbal items in the corresponding test form.

10.2.1   Preparing the ADM Data for kequate

The NEAT design produces bivariate score data, where the score vectors contain pairs of the total test scores and total anchor test scores. If the files are placed in the current working directory, the function load() can be used to load the data. After loading the data and before creating frequency distributions, we need to first create vectors of the total verbal test scores on test forms X and Y and the total verbal anchor scores on the anchor test form A. In the two data sets, the anchor items are the first 40 items, and the unique items are the next 80 items in each of the test forms. The following code can be used to perform these tasks:

```
> load("ADMX.Rda")
> load("ADMY.Rda")
> verb.xa <- apply(ADMX[,1:40],1,sum)
> verb.ya <- apply(ADMY[,1:40],1,sum)
> verb.x  <- apply(ADMX[,41:120],1,sum)
> verb.y  <- apply(ADMY[,41:120],1,sum)
```

Next, the kefreq() function from **kequate** is used to produce bivariate score frequency distributions that are needed under the NEAT design. The kefreq() function has the arguments in1, which receives a numerical vector with the total test scores on test X, and in2, which receives a numerical vector with the total test scores on either test Y, when a SG design is being considered, or the anchor test scores A, as in this case. The xscores and ascores are used to indicate the score scales for X and A, respectively.

```
> neat.x <- kefreq(in1 = verb.x, xscores = 0:80,
+ in2 = verb.xa, ascores = 0:40)
> neat.y <- kefreq(in1 = verb.y, xscores = 0:80,
+ in2 = verb.ya, ascores = 0:40)
```

The obtained objects contain the different combinations of test scores together with the frequencies of the combinations. We could, for example, choose to inspect two combinations of test scores by printing lines 1000 and 1001 of the frequency table, as follows:

```
> neat.x[1000:1001,1:3]
     X  A frequency
1000 27 12         3
1001 28 12        10
```

Line 1000 indicates that there were three test takers who scored 27 on test form X and 12 on the anchor test form A. Similarly, line 1001 indicates that there were 10 test takers who scored 28 on test form X and 12 on the anchor test form A, respectively.

The marginal frequencies of X and A in population $\mathcal{P}$, and of Y and A in population $\mathcal{Q}$, can be graphically illustrated as in Figure 10.1. The **equate** package was used to create this figure as it automatically generates it when a `freqtab` object is passed to the `plot()` function. The function `freqtab()` was used to create bivariate frequency distributions for population $\mathcal{P}$ with the objects `verb.x` and `verb.xa`, and for population $\mathcal{Q}$ with the objects `verb.y` and `verb.ya`. The following code was used to produce the plots of the bivariate score distributions for test form X and the anchor test form A, and for test form Y and the anchor test form A.

```
> library(equate)
> feqX <- freqtab(cbind(verb.x,verb.xa),scales=list(0:80,0:40))
> feqY <- freqtab(cbind(verb.y,verb.ya),scales=list(0:80,0:40))
> plot(feqX,xlab="Test form X",ylab="Anchor test form A")
> plot(feqY,xlab="Test form Y",ylab="Anchor test form A")
```

10.3   Simulated Polytomous Data

We will illustrate how to perform IRT kernel equating using simulated polytomous data under the NEAT design. The simulated data consider two different test forms, each containing 40 unique items, and an external anchor test form with 15 anchor items. Possible item scores are 0, 1, or 2, and the GPCM given in Eq. (3.20) is used to model the items.

To construct the simulated data we proceeded as follows. First, two threshold parameters and one discrimination parameter are generated for each item. In the code below, the first two entries in each list stored in the objects Xp and Yp correspond to the threshold parameters, and the third entry contains the item discrimination parameter.

To obtain randomly generated response data for $1,000$ test takers, we used the `rmvordlogis()` function from the **ltm** package. The first argument

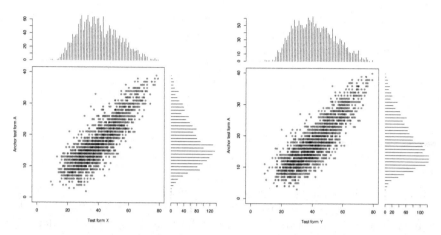

FIGURE 10.1: Marginal frequencies of X and A (left panel) and marginal frequencies of Y and A (right panel) for the ADM data.

indicates the number of response patterns to simulate (1000), the second argument states which data to use (Xp,Yp), the third argument indicates from which model to simulate, in our case the GPCM (gpcm), and the fourth argument is a numeric vector providing the values of the latent trait to be used in the simulation (z.vals). The rmvordlogis() returns response data for each of the items with the following item scores, 1, 2, and 3. However, as we wanted the item scores to be 0, 1, and 2, we subtracted 1 from the simulated response data (XsP, YsP). Thus, to generate $1,000$ test takers' response patterns with the GPCM for the test forms, we used the following code.

```
> Xp <- vector("list", 55)
> Yp <- vector("list", 55)
> set.seed(7)
> for(i in 1:55)
> Xp[[i]] <- c(rnorm(1,-0.2,1),rnorm(1,0.4,1),runif(1))
> set.seed(9)
> for(i in 1:55)
> Yp[[i]] <- c(rnorm(1,-0.2,1),rnorm(1,0.4,1),runif(1))

> library(ltm)
> XsP <- rmvordlogis(1000,Xp,model="gpcm",z.vals = rnorm(1000))
> YsP <- rmvordlogis(1000,Yp,model="gpcm",z.vals = rnorm(1000))
> colnames(XsP) <- paste("PX",1:55)
> colnames(YsP) <- paste("PY",1:55)
> XoP <- XsP-1
> YoP <- YsP-1
```

As we will use **mirt** (Chalmers, 2012) for the IRT analyses, and since that package requires unique item names, we added headers to the item columns using the `colnames()` function in the code above, so that the items are labeled as `PX1,...,PX55` and `PY1,...,PY55`. The items `PX41,...,PX55` and `PY41,...,PY55` are used as the external anchor test form.

10.4   Step 1: Presmoothing

There are at least four possible ways to conduct the presmoothing step, as seen in Chapter 3. We could use polynomial log-linear models (Section 3.2.1), nonparametric discrete kernel estimators (Section 3.3.1), beta4 models (Section 3.4.1), or IRT models (Section 3.4.2). In this chapter, we will illustrate presmoothing with polynomial log-linear models and presmoothing with IRT models. As noted in Chapter 3, we can use either chained equating (CE) or poststratification equating (PSE) under a NEAT design, and both approaches will be used in this chapter. Note that previously in this book, we have referred to these as NEAT CE and NEAT PSE. However, since this entire chapter is devoted to the NEAT design, we will frequently abbreviate them and just use CE and PSE.

We use the R package **kequate** to illustrate presmoothing with log-linear modeling. Note that it is also possible to use the **R** package **SNSequate** to presmooth with log-linear models. For examples with **R** code using that package, refer to Chapter 4 in González and Wiberg (2017).

10.4.1   Log-Linear Models

Test score frequencies are count data that can be assumed to follow a Poisson distribution. The general `glm()` function in **R** can thus be used to fit log-linear models to the score frequencies by setting the argument `family="poisson"`. The estimated score probabilities will be obtained differently depending on the considered data collection design. For instance, under the EG design, we can divide the obtained fitted values $\hat{n}_j$ by the sum of the frequencies and then get estimates of r_j and s_k directly. In the NEAT design, however, we need to use the design function to transform the vector of probabilities in order to obtain $\hat{r}_j$ and $\hat{s}_k$. Once a `glm()` object has been created, the **kequate** function receives it as an input and internally performs all the needed calculations to obtain estimated score probabilities under any of the designs included in the GKE framework. When using the `glm()` to fit log-linear models, we do not need to state at this point whether we plan to use CE or PSE, as the `kequate()` function will automatically account for the different methods to estimate the score probabilities. The focus in this section is thus on how to use the `glm()` function to estimate polynomial log-linear models.

Suppose we want to fit bivariate log-linear presmoothing models as described in Eq. (3.5). To fit score data from test forms X and A, we set $T_r = 3$ and $T_a = 1$, and from test forms Y and A, we set $T_r = 2$ and $T_a = 1$. These models can be fitted in **R** by using the previously created objects `neat.x` and `neat.y` using the following code.

```
> NEATvX <- glm(frequency~I(X)+I(X^2)+I(X^3)+I(A)+I(X):I(A),
+ family = "poisson", data = neat.x, x = TRUE)
> NEATvY <- glm(frequency~I(X)+I(X^2)+I(A)+I(X):I(A),
+ family = "poisson", data = neat.y, x = TRUE)
```

Model fit can be evaluated from the output by typing `summary(NEATvX)`, which yields the following output:

```
Call:
glm(formula = frequency ~ I(X)+I(X^2)+I(X^3)+I(A)+I(X):I(A),
    family = "poisson", data = neat.x, x = TRUE)


Coefficients:
              Estimate Std. Error z value Pr(>|z|)
(Intercept) -4.737e+00  3.637e-01 -13.025  < 2e-16 ***
I(X)         4.214e-01  2.812e-02  14.986  < 2e-16 ***
I(X^2)      -8.569e-03  6.832e-04 -12.542  < 2e-16 ***
I(X^3)       3.724e-05  5.167e-06   7.208 5.66e-13 ***
I(A)        -1.792e-01  7.828e-03 -22.896  < 2e-16 ***
I(X):I(A)    3.976e-03  1.837e-04  21.637  < 2e-16 ***
---
Signif. codes:  0 '***' 0.001 '**' 0.01 '*' 0.05 '.' 0.1 ' ' 1

(Dispersion parameter for poisson family taken to be 1)

    Null deviance: 7266.1  on 3320  degrees of freedom
Residual deviance: 4871.7  on 3315  degrees of freedom
AIC: 6786.9

Number of Fisher Scoring iterations: 6
```

From this output, we can see that all variables included in the model are significant, i.e., they are relevant for modeling our data. We can also see the model AIC and deviance values. These values should be low in comparison to the values of other competing models. The number of Fisher scoring iterations refers to the number of iterations needed until convergence of the algorithm used to obtain the model fit. Note that the chosen models shown here are only illustrative, so it is important to fit several log-linear models and compare them, for instance with respect to AIC and deviance, in order to decide which model fits the data best. How to build more complex models and how to assess the model fit are discussed next.

10.4.1.1　Modeling Complexities

The observed distributions might be complex in several ways. Two common cases are discussed here, and can also be found in von Davier et al. (2004, Chapter 10) and in González and Wiberg (2017, Chapter 4). First, because scores are most of the time rounded to integer values, there may be "gaps" in the observed score frequencies that occur at regular intervals. Second, there might be a "spike" at 0 in the score distribution because negative values have been rounded to 0. Negative score values can occur in a multiple-choice test if a fraction of the total test score is discounted for every incorrect answer given by the test taker. von Davier et al. (2004) suggested creating indicator variables to account for the particular gaps and spikes values, which are included in the log-linear model.

To illustrate how complexities in the data can be modeled when using the `glm()` function, we assigned `neat.x` to the object `neat.xm`, which we manipulated to create a score distribution which exhibits a spike at $X = 0$, with a frequency of 10 test takers whose negative scores were truncated; and gaps at $X = 35, 40$ where the frequency of scores is lower than expected (20 in this case). The following code was used to obtain this data.

```
> neat.xm <- neat.x
> neat.xm$frequency[neat.xm$X==0]   <- 10
> neat.xm$frequency[neat.xm$X==35] <- 20
> neat.xm$frequency[neat.xm$X==40] <- 20
```

Next, using the code shown below, we created two indicator variables to account for these irregularities; `ix1` that takes the value 1 if X is equal to 0, and `ix2` that takes the value 1 if X takes values in the score set $\{35, 40\}$ where X exhibits gaps:

```
> neat.xm$ix1 <- numeric(length(neat.xm$X))
> neat.xm$ix2 <- numeric(length(neat.xm$X))
> neat.xm$ix1[neat.xm$X==0] <- 1
> neat.xm$ix2[neat.xm$X %in% c(35,40)] <- 1
```

Finally, we fitted two models to the manipulated data, one where no indicator variables are included (`niX`), and another one that accounts for the irregularities including the indicator variables (`iX`), using the following code.

```
> niX <- glm(frequency~I(X)+I(X^2)+I(X^3)+I(A)
+ I(X):I(A),data = neat.xm,
+ family = "poisson", x = TRUE)

> iX <- glm(frequency~I(X)+I(X^2)+I(X^3)+I(A)
+ I(X):I(A)+I(ix1)+I(ix2),data = neat.xm,
+ family = "poisson", x = TRUE)
```

Typing `summary(niX)` and `summary(iX)` (not shown here), it can be seen that the model fit is greatly improved when the indicator variables are included in the model. In particular, the AIC value is reduced from 14106 to 7473.1, and all the predictor variables in the latter model were statistically significant.

To include all possible complexities in the data, we can create additional variables for particular values of the X, Y, and/or A variables.

10.4.1.2 Log-Linear Model Fit

As discussed in Section 3.5.1, there are a number of possibilities for evaluating the model fit of log-linear models. When using the `glm()` function to fit log-linear models, we have seen that the output provides the AIC measure by default. The BIC measure can also be obtained using the function `BIC()`. For instance, the BIC for the log-linear model fitted in the preceding section and stored in the object `iX` is 7521.941 and can be retrieved typing `BIC(iX)`. Both the AIC and BIC measures should be compared for competing models, and we should choose the model with the smallest values of AIC and/or BIC.

As an example, additionally to the model fitted in Section 10.4.1 which was stored in the object `NEATvX`, we can fit three additional log-linear models with different polynomial degrees and interactions between the variables, and compare them in term of both AIC and BIC. The following code can be used:

```
> NEATvX.2 <- glm(frequency~I(X)+I(X^2)+I(X^3)+I(A)+I(A^2)
+ I(X):I(A^2), family = "poisson",
+ data = neat.x, x = TRUE)

> NEATvX.3 <- glm(frequency~I(X)+I(X^2)+I(A)+I(A^2)
+ I(X):I(A^2), family = "poisson",
+ data = neat.x, x = TRUE)

> NEATvX.4 <- glm(frequency~I(X)+I(X^2)+I(A)+I(A^2)
+ I(X^2):I(A^2), family = "poisson",
+ data = neat.x, x = TRUE)

> aic.com<-AIC(NEATvX,NEATvX.2,NEATvX.3,NEATvX.4)
> bic.com<-BIC(NEATvX,NEATvX.2,NEATvX.3,NEATvX.4)

> cbind(aic.com,bic.com)
          df      AIC df      BIC
NEATvX     6 6786.942  6 6823.590
NEATvX.2   7 3713.685  7 3756.441
NEATvX.3   6 3824.459  6 3861.107
NEATvX.4   6 4350.068  6 4386.716
```

According to both the AIC and BIC values, the best fitting model is `NEATvX.2`. It should be noted that the AIC and BIC values may contradict each other, which was not the case in this example. An alternative option is to compare different models with the likelihood ratio χ^2 test. For more details and **R** code for using these measures in **R**, refer to González and Wiberg (2017) Section 2.3.4.

Another possible strategy mentioned in Section 3.5.1 is to assess conditional parameters such as conditional means, variances, skewnesses, and kurtoses. The chosen log-linear model is appropriate if the conditional parameters do not differ too much between the observed and estimated distributions. The cdist() function in **kequate** can be used to obtain the conditional moments for the observed and estimated bivariate frequency distributions, which is illustrated next. The cdist() function takes as arguments the matrix of estimated (est) bivariate score probabilities, the matrix of observed (obs) bivariate score probabilities, and the possible score values for the two test forms. To obtain data frames with the conditional parameters for the observed and estimated bivariate frequency distributions, as well as to plot them, we can use the following code:

```
> obsNEATx <- matrix(neat.x$freq, nrow=81)/sum(NEATvX$y)
> estNEATx <- matrix(NEATvX$fitted.values,nrow=81)/sum(NEATvX$y)
> distNEATx <- cdist(est=estNEATx, obs=obsNEATx, 0:80, 0:40)
> plot(distNEATx)
```

In this code, nrow refers to the number of different test scores, the observed frequencies are obtained from the object neat.x, and the fitted values are obtained from the object NEATvX. Finally, 0:80 represents the possible scores on test form X and 0:40 represents the possible scores on the anchor test form A.

Figure 10.2 shows the plots of the observed and estimated conditional mean and variances for the distributions $X \mid A$ and $A \mid X$. The estimated values are represented by squares and the observed values are represented by triangles. Note that the labels in this figure are the ones that **kequate** uses by default. Here, "Score values test X" refers to the score values of test form X, and "Score values test A/Y" refers in this case to the score values of the anchor test form A. Further, "Mean of $X \mid A$" refers to the conditional expected score X given A and "Mean of $A \mid X$" refers to the conditional expected score A given X. Similar definitions apply for the conditional variances. The examined log-linear model will be appropriate if the conditional parameters of the estimated distribution do not deviate too much from the observed distribution. Not surprisingly, for this example this is not the case as NEATvX presented the worst fitting in terms of both AIC and BIC among all the fitted models. To illustrate that a better model would exhibit more attenuated differences between the observed and the fitted values, Figure 10.3 shows the same type of plots for model NEATvX.2.

From this figure, it can be seen that, indeed, differences have been attenuated, but there is still room for other models that could fit the data even

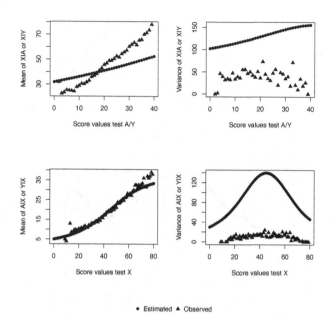

FIGURE 10.2: Conditional mean and variances of $X \mid A$ and $A \mid X$ for model NEATvX.

better. Note that these plots were used for illustration and several models should be examined in order to find the best-fitting model.

10.4.2 Binary IRT Models

We can use either the **mirt** or the **ltm** **R** packages to fit IRT models for binary-scored data, whose output will be used to obtain a smoothed version of the score probability distributions. For an example of presmoothing using **R** code where the 2PL IRT model is fitted with the **ltm** package, refer to Section 7.3.3 in González and Wiberg (2017). To fit an IRT model with the **mirt** package we use the function `mirt()`, which takes as first argument a matrix or a data frame of the item data. The second argument states the number of factors to extract (`model`), i.e., the number of underlying dimensions, which for a unidimensional model is 1. The third argument states which IRT model to fit (`itemtype`), and the final argument is a logical argument, which states whether we want standard errors of the item parameters (`SE`) to be computed. To fit a 2PL IRT model to the two test forms X and Y and the anchor test form A of the ADM data, we can thus write as follows:

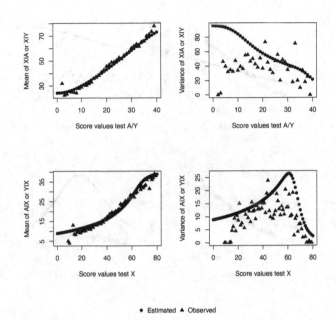

FIGURE 10.3: Conditional mean and variances of $X \mid A$ and $A \mid X$ for model NEATvX.2.

```
> library(mirt)
> ADMx.2PL<-mirt(data.frame(ADMX),model=1,itemtype="2PL",SE=TRUE)
> ADMy.2PL<-mirt(data.frame(ADMY),model=1,itemtype="2PL",SE=TRUE)
```

The mirt objects ADMx.2PL and ADMy.2PL contain the item parameter estimates, and these objects can be used as inputs in **kequate** to perform the actual equating. The parameter estimates for the first item can be viewed by writing.

```
> head(coef(ADMx.2PL),1)
```

The resulting output is given below, where the item parameter estimates are on the first row. The second and third rows give the limits of a 95% confidence interval for the parameters. By default, the **mirt** output shows values for item parameters as if a four-parameter IRT model was fitted. In the output below, a1 represents the item discrimination, d represents the item difficulty, g represents the estimated guessing parameter value (or the lower asymptote), and u is the upper asymptote. As we have estimated a 2PL IRT model, the last two parameters are set to 0 and 1, respectively, and this is why no confidence intervals are available for them.

	a1	d	g	u
par	1.520613	0.123072792	0	1
CI_2.5	1.357217	-0.002024216	NA	NA
CI_97.5	1.684008	0.248169799	NA	NA

10.4.2.1 IRT Model Fit

Before proceeding to use an IRT model for presmoothing, it is essential to check whether the IRT assumptions are fulfilled, and if the model fits the data (see Section 3.5.4). In this section, we will briefly describe how to examine unidimensionality, local independence, parameter invariance, and monotonicity in the context of IRT models. We will also show how to assess model goodness of fit, item fit, and person fit of the IRT models.

The unidimensionality assumption can be checked by, for example, performing an exploratory factor analysis, in which case the assumption is fulfilled if only one clear factor is visible, see Section 3.5.4. Exploratory factor analysis and the examination of the dimensionality can be performed with the **R** package **psych**. Graphically, to examine if only one dominant factor is present, we can inspect a scree plot (**scree**). We can also use non-graphical tools, for instance, the VSS criterion (**vss()**), or the MAP criterion (which is included in the **vss()** function), to examine unidimensionality. The VSS criterion assesses how well the factor structure aligns with a simple pattern characterized by high loadings of variables on a small number of factors, and low or negligible loadings on other factors. The MAP criterion assesses the adequacy of different factor solutions by considering the average squared partial correlations between the variables, which are a measure of the residual variance left unexplained by the extracted factors. For other measures that can be used to examine dimensionality together with **R** examples, refer to the documentation of **psych**. To be able to use these measures with our data, we first construct a data matrix and select the 80 items of test form X that are of interest.

```
> admP1  <- as.matrix(ADMX[,c(41:120)])
```

To examine unidimensionality in test form X, we choose to use a scree plot and the **vss()** function using the following code.

```
> scree(admP1,factors=TRUE,pc=FALSE)
> vss(admP1,3)
```

The first line of code results in a scree plot of the eigenvalues from a factor analysis (**factors=TRUE**) of the selected items and where we have indicated that we do not want to show a principal component solution (**pc=FALSE**). The resulting plot can be seen in Figure 10.4. From this scree plot, we can see that there is one factor with a large eigenvalue and the other factors have smaller associated eigenvalues, and it is thus reasonable to assume unidimensionality in this case. The second line of code calculates the VSS criterion. The 3 in

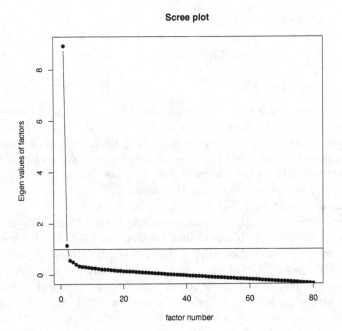

FIGURE 10.4: Scree plot for test form X.

the vss() means number of factors to extract, and it should always be larger than hypothesized. Note that the example code is only used for illustrative purposes, and better models might be possible to obtain. The output of the vss() function is the following:

```
Very Simple Structure
Call: vss(x = admP1, n = 3)
VSS complexity 1 achieves a maximimum of 0.58  with  1 factors
VSS complexity 2 achieves a maximimum of 0.6   with  2 factors

The Velicer MAP achieves a minimum of 0  with  2  factors
BIC achieves a minimum of  -18634.28  with  2  factors
Sample Size adjusted BIC achieves a minimum of  -9099.94
with  2  factors

Statistics by number of factors
  vss1 vss2    map dof chisq    prob sqresid  fit RMSEA    BIC
1 0.58 0.00 0.0011 3080  5229 3.1e-115     68 0.58 0.019 -18182
2 0.36 0.60 0.0011 3001  4176  3.8e-42     65 0.60 0.014 -18634
3 0.28 0.54 0.0012 2923  3831  1.1e-27     63 0.61 0.013 -18386
```

	SABIC	complex	eChisq	SRMR	eCRMS	eBIC
1	-8396	1.0	9244	0.027	0.027	-14167
2	-9100	1.6	6523	0.023	0.023	-16288
3	-9099	2.0	5827	0.021	0.022	-16391

This output states that according to the VSS criterion, either a one-factor solution or a two-factor solution is preferable. When using the vss() function, we also get information about the MAP criterion, which in this case suggests that a two-factor solution is preferable. A plot of the VSS criterion is also available by default but was excluded here. Note that it is not uncommon that the VSS criterion and MAP criterion yield different solutions of number of factors to be used. In this case, we choose a one-factor solution based on the scree plot and that it was one of the suggestions according to the VSS criterion. The choice of a one-factor solution means that we conclude that unidimensionality is fulfilled.

The assumption of local independence can be qualitatively checked by examining the content of the different items to make sure that one item does not give the answer to another. Quantitative measures such as Chi-square tests, as well as different statistics described by Chen and Thissen (1997) can also be used.

In the **mirt** package, we can examine the residuals from the IRT models to examine local independence. In the code below, we have used the option Q3, which implements the correlation magnitude Q_3 proposed by Yen (1984a) and described in Chen and Thissen (1997) to examine local dependence. An alternative to Q3 is to set the option "LD" which uses a chi-square statistic, as described in Chen and Thissen (1997). The argument **suppress** is a numeric value indicating which pair of local dependency combinations of items to flag as being too high. Thus, only absolute values of the estimates greater than this chosen value will be returned, while all values less than this value will be set to NA. In the example code below, we set this numerical value to 0.2, as that is the suggested critical value by Chen and Thissen (1997). The code for displaying whether the first five items exhibited local dependency with any of the other items in the test form using the item model residuals is as follows:

```
> head(residuals(ADMx.2PL,"Q3",suppress=0.2),5)
```

The resulting output is a table of summary statistics (e.g., mean, min, max) as well as a large matrix containing the correlation magnitude for each pair of items, which we have excluded here due to its size. For these five items, all item pairs were flagged with NA. Moreover, the summary statistics show that values of Q_3 for all item pairs are in the range −0.139 - 0.189 with a mean value −0.007, so we concluded that local independence is fulfilled.

Parameter invariance can be checked by examining if the parameter estimates are stable when they are estimated using different groups. This means that we can estimate the IRT models first by using, for example, only the girls who answered the items, and then repeat the analysis with only the boys who answered the items. If the obtained estimates are similar, this assumption is fulfilled. As it is straightforward to conduct such analyses, we omit an example here.

To examine model fit, one should examine item fit and person fit with different infit and outfit measures. Outfit measures are sensitive to unexpected observations by persons or items. The infit statistics are sensitive to unexpected patterns of observations of persons or items that are aimed to target them. To obtain item fit measures, we can use `itemfit()` function in the **mirt** package as follows:

```
> itemfit(ADMx.2PL)
      item    S_X2 df.S_X2 RMSEA.S_X2 p.S_X2
...
42  V1X02  81.116      66      0.011  0.100
43  V1X03  62.112      72      0.000  0.791
...
```

For the item fit, we have suppressed the output for all items except for items 2 (`V1X02`) and 3 (`V1X03`) in test form X. The output from the `itemfit()` function yields signed chi-square tests (Orlando and Thissen, 2000) (`S_X2`) with degree of freedom (`df.S_X2`) and the p-values (`p.S_X2`) together with RMSEA for each item. The chi-square test has the null hypothesis that the model fits and thus should be nonsignificant, and RMSEA values below 0.06 are indications of good item fit (Hu and Bentler, 1999). Both these items were concluded to show good item fit, which was also true for most of the other items.

To examine person fit for each test taker, we can use the `personfit()` function in the **mirt** package. The following code yields the person fit for the first two test takers.

```
> head(personfit(ADMx.2PL),2)
      outfit    z.outfit    infit   z.infit         Zh
1 0.9896358 -0.02918623 1.056966 0.6494214 -0.4037369
2 1.1901317  0.87777501 1.035300 0.3586948 -0.5233393
```

The output shows outfit and infit measures together with the Z_h values (Drasgow et al., 1985). The Z_h values are the standardized maximum likelihood indices for each person, as described in Section 3.5.4. The empirical distribution of Z_h is close to a standard normal distribution. A good fit has a value close to 0, and person misfit can be concluded if the value is further away than two standard errors. In this example, both test takers have good person fit as the Z_h values are relatively close to zero. Figure 10.5 shows the Z_h values for all test takers in the sample. As most of the Z_h values in this figure lie between the range -2 to 2, this leads to the conclusion that there is a good fit.

Once the assumptions and model fit have been checked, the IRT model can be safely used to obtain presmoothed score distributions. It is, however,

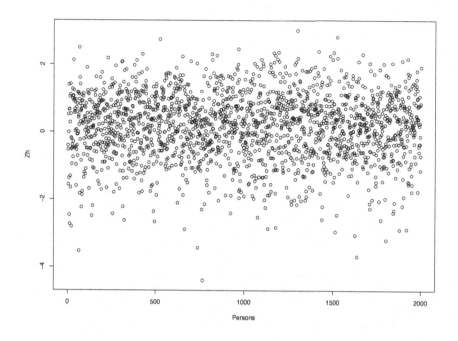

FIGURE 10.5: Z_h values for test takers administered test form X.

also recommended to inspect the item characteristic curves (ICC), i.e., the graphical representation of the response function (i.e., the probability of answering an item correctly depending on the test taker's ability) for each item. The ICC is used to examine the properties of the items and to easily compare different items. We can also use the ICC to graphically check the assumption of monotonicity. Monotonicity is concluded if the ICC exhibit a consistent pattern of increasing probability as the latent ability level increases.

Two examples of ICCs are shown in Figure 10.6 for item 2 (solid curve) and item 3 (dotted curve). The figure was obtained using the code below, where we first fitted a two-parameter IRT model with the **ltm** package. In this example, item 2 is more discriminating (steep curve) than item 3 (curve is relatively flat), and item 3 is easy for many test takers, while item 2 is especially easy for test takers with high ability. Also, from inspection of this plot, the assumption of monotonicity is concluded to be fulfilled for these items as the probability of answering the items correctly increases with the latent ability level for both items.

```
library(ltm)
> admP    <- as.matrix(ADMX[,c(41:120,1:40)])
> adm2p1P <- ltm(admP~z1, IRT.param=TRUE)
```

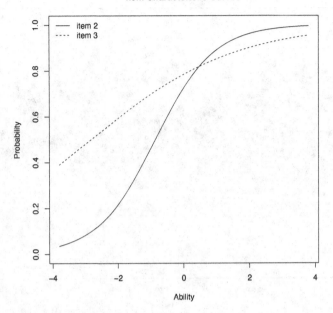

FIGURE 10.6: ICCs for items 2 and 3 in test form X.

```
> plot(adm2plP,items = c(2,3),lty=c(1,2),col=c(1,1),
+ legend = TRUE, labels=c("item 2","item 3"))
```

If we have competing IRT models, we can compare the log-likelihood, AIC, and BIC for different models. A model with lower AIC and/or lower BIC indicates a better model fit. These three measures can be easily obtained by using the `extract.mirt()` function, which retrieves internal objects from estimated models, in this case the log likelihood, AIC, and BIC, respectively.

```
> extract.mirt(ADMx.2PL,"logLik")
[1] -141394.3
> extract.mirt(ADMx.2PL,"AIC")
> [1] 283268.6
> extract.mirt(ADMx.2PL,"BIC")
[1] 284612.8
```

To compare the fit of two different models, for example the Rasch model and the 2PL IRT model, we can use the `anova()` function. In the code below, we first fit a Rasch model and then, in the second row, we compare the Rasch model with the 2PL IRT model that was previously fitted.

```
> ADMx.R <- mirt(data.frame(ADMX),1,itemtype="Rasch",SE=TRUE)
> anova(ADMx.R,ADMx.2PL)
```

From the output (shown below), the chi-square test (X2) and its resulting p-value (p) allow us to conclude that there is a significant difference between the models that favor the 2PL model. This conclusion can be confirmed based on the AIC and BIC values, which are smaller for the 2PL IRT model, and thus this model is preferable over the Rasch model.

```
            AIC    SABIC       HQ      BIC    logLik     X2 df p
ADMx.R   286157.2 286450.5 286406.0 286834.9 -142957.6
ADMx.2PL 283268.6 283850.4 283762.2 284612.8 -141394.3 3126.5 119 0
```

Summing up, for the binary-scored item data, the IRT assumptions were concluded to be fulfilled, and the chosen 2PL model resulted in a reasonable item fit and person fit. The 2PL model can thus be used in the presmoothing step to obtain smoothed estimated score probabilities, which will be illustrated in Section 10.5.2.

10.4.3   Polytomous IRT Models

To illustrate presmoothing with polytomously scored items for the NEAT design, we use the simulated data example described in Section 10.3. To presmooth polytomously scored item data with IRT models, we can use either the **mirt** or the **ltm R** packages. Here we used the mirt() function from the **mirt** package to estimate the item and person parameters with the following arguments. We specified the data used (XoP,YoP), which model should be estimated for each item in the data (itemtype), which in this case is the GPCM (gpcm), and indicated that we wanted standard errors to be computed (SE=T).

```
> library(mirt)
> X55r <- mirt(XoP,1,itemtype=rep('gpcm',55),SE=T)
> Y55r <- mirt(YoP,1,itemtype=rep('gpcm',55),SE=T)
```

Each obtained object (X55r,Y55r) contains estimated item parameters, using the GPCM, for the 40 unique items and the 15 anchor items. The estimated item parameters for the X and A test forms administered to population $\mathcal{P}$ can be seen by typing coef(X55r,IRTpars=TRUE,simplify=TRUE). In the output below, the parameter estimates for the first five items are displayed, where a is the item discrimination parameter, and b1 and b2 are the threshold parameters.

```
          a      b1      b2
PX 1   0.203   2.254  -1.340
PX 2   0.461   0.949   2.233
PX 3   0.430  -1.160   1.183
PX 4   0.330  -1.176  -2.100
PX 5   1.147   0.170   2.735
```

10.4.3.1 IRT Model Fit

Similar to binary-scored IRT models, before proceeding to use polytomous IRT models we need to examine the IRT assumptions and model fit, using the measures discussed in Section 3.5.4. In this section, we will briefly describe how to examine unidimensionality, local independence, parameter invariance, monotonicity, and goodness of fit in terms of item fit and person fit of the IRT models.

Unidimensionality can be examined in a way similar to the binary-scored IRT models, but here we choose to also use parallel analysis and factor analysis. In brief, in parallel analysis, factors are extracted until the eigenvalues of the observed data matrix are smaller than the corresponding eigenvalues of a randomly chosen data set of the same size. To examine unidimensionality we can, for example, use the following code.

```
> XoPm <- as.matrix(XoP[,c(1:40)])
> scree(XoPm,factors=TRUE,pc=FALSE)
> vss(XoPm,3)
> fa.parallel(XoPm,nfactors=1)
```

As the first three lines of code above are similar to the code used for binary-scored data (except that we use XoPm instead of admP), we have omitted a more elaborated discussion here. We have, however, in the last line included how to perform a parallel analysis. The function fa.parallel() takes as input the data matrix of interest (XoPm) and the number of factors we wish to examine. The output is given below, and it simply suggests that we have a one factor solution. Note that we excluded the scree plot which was also an outcome from this function, as we have already shown an example of a scree plot for the binary-scored data.

```
Parallel analysis suggests that the number of factors =  1
and the number of components =  1
```

As we have already described and discussed how to check for unidimensionality using the other measures when we had binary-scored items, the outputs are excluded here. We have, however, also included an example when using factor analysis to decide the number of factors. In this example we examine the possibility of using one or two factors (nfactors=2) by writing:

```
> fa(XoPm,nfactors=2)
```

The resulting output yields factor loadings and different model fit measures. Here we show the part of the output where the proportion of variance explained using one and two factors is given.

```
                 MR1   MR2
SS loadings      5.43 0.46
```

```
Proportion Var          0.14 0.01
Cumulative Var          0.14 0.15
Proportion Explained  0.92 0.08
Cumulative Proportion 0.92 1.00
```

In this case, the proportion of variance explained for the first factor is 0.92 and only 0.08 for the second factor, and thus we conclude that a one factor solution is suitable. Note that in this example, the scree plot, the VSS criterion, the MAP criterion, the parallel analysis, and the factor analysis all suggested a one factor solution; i.e., we conclude that the unidimensionality assumption is fulfilled.

Parameter invariance can be examined by checking if the parameter estimates are stable when they are estimated using different groups. If the obtained estimates are similar, this assumption is fulfilled. We will not include such analyses here as it is not difficult to run the same analyses for different groups and compare if the results are similar.

Local independence between the items can be examined with the Q_3 measure described in Section 10.4.2.1. For the polytomous scored items, we use the following code to examine local independence for test form X.

```
> X40r <- mirt(XoPm,1,itemtype=rep('gpcm',40),SE=T)
> residuals(X40r,"Q3",suppress=0.2)
```

The output is excluded as it is a large matrix containing the Q_3 values for all the item pairs. "NA" in the matrix entries indicates local independence between the items, which was the case for all the items. Also, the values of Q_3 ranged from -0.131 to 0.096 with a mean value -0.019, which confirms that local independence is fulfilled.

Similarly as for the binary-scored items, we can evaluate model fit by examining both item fit using the `itemfit()` function and person fit for each test taker using the `personfit()` function from the **mirt** package. These two functions can be used with our example by simply writing

```
> itemfit(X55r)
> personfit(X55r)
```

The output for the item fit is similar to the one for binary-scored items as it includes chi-square tests together with RMSEA values for each item, as seen below for the first five items.

```
    item      S_X2 df.S_X2 RMSEA.S_X2 p.S_X2
1   PX 1  124.900     125      0.000   0.486
2   PX 2  117.712     104      0.011   0.169
3   PX 3   97.850     113      0.000   0.844
4   PX 4  121.304     117      0.006   0.374
5   PX 5   77.304      74      0.007   0.374
```

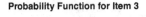

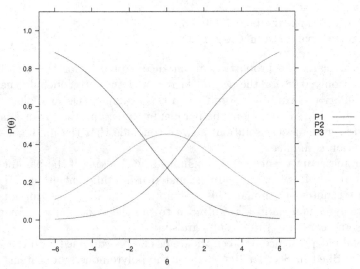

FIGURE 10.7: Item plot for item 3.

These five items were concluded to have good fit based on the fact that all chi-square tests are nonsignificant and all the RMSEA values are below 0.06. Note that the other items, for which the output was suppressed here, can also be concluded to have good item fit.

When using polytomously scored items, the output for person fit shows the response pattern for each test taker together with outfit and infit measures and the Z_h values, which was also used with binary-scored item data in Section 10.4.2.1. The person fit output for the first two test takers is shown below with two decimals (only the answers to the first four items are shown).

```
 PX.1 PX.2 PX.3 PX.4 PX.5 outfit  z.outfit infit z.infit  Zh
1   0    0    1    2    0    1.04    0.36    1.04   0.35  -0.17
2   0    0    0    2    0    1.14    1.11    1.14   1.15  -0.33
```

Recall that the empirical distribution of Z_h is close to a standard normal distribution so that person misfit can be concluded if this value is further away than two standard errors. In this example, the first two test takers can be concluded to have good person fit according to the Z_h value. A plot of the Z_h values for all the test takers (not shown) shows that most of the values are in the expected range, -2 to 2, which allows us to conclude good fit.

Finally, for ordered polytomous IRT models, the assumption of monotonicity should hold (Kang et al., 2018). For GPCM, the monotonicity assumption means that category thresholds are ordered correctly according to the latent trait. Thus the first response category should have a higher probability for test takers with low latent ability, the second response curve should have a higher probability for test takers with mid ability, etc. To examine monotonicity, we can plot the ICCs of specific items by using `itemplot()`, in which we indicate which item we want to plot. In the code below, we plot item 3, which is displayed in Figure 10.7, where the probability curves for obtaining each response alternative of the item are shown. In this plot, we can see that test takers scoring in the first category (P1 in the plot) are in general those having low values of the latent trait, those scoring in the mid category (P2 in the plot) are of mid latent trait, and those scoring in the highest category (P3 in the plot) have a higher value of the latent trait. Therefore the assumption of monotonicity is concluded to be fulfilled for this item.

```
> itemplot(X55r,3)
```

Summing up, for the polytomously scored data, all the IRT assumptions were concluded to be fulfilled, and the chosen IRT models showed good item fit and person fit. This means that we can use the chosen IRT models for presmoothing and proceed to the next step, in which the score probabilities are estimated.

10.5 Step 2: Estimating Score Probabilities

When performing GKE in **kequate**, we use either the function `kequate()` or the function `irtose()`, as described in Section 10.7 and in Section 10.7.2, depending on which model has been fitted in the presmoothing step. For instance, if log-linear models are used in the presmoothing step, the `glm` objects are fed into `kequate()`, whereas if IRT models are used, then the corresponding objects are fed into `irtose()` as input. Both functions automatically calculate and store the estimated score probabilities that are used later in the equating step.

In the example that follows, we assume that equating is conducted under the NEAT-PSE design. To obtain the estimated score probabilities, we will use the `getScores()` function in **kequate**. This function receives as input an equating object created with either `kequate()` or `irtose()`, and returns the estimated score probabilities according to the equating design considered.

10.5.1 Log-Linear Models

When using log-linear models in the presmoothing step, an equating object will be created in Section 10.7 and will be called `glmPSE`. The estimated score probabilities can be obtained and stored in the objects `rj` and `sk` by simply writing the following code:

```
> rj <- getScores(glmPSE)$X$r
> sk <- getScores(glmPSE)$Y$s
> cbind(0:80,rj,sk)
```

The resulting output of the estimated score probabilities for the first five score values is as follows.

```
              rj           sk
[1,]   0 2.661053e-05 0.0002230999
[2,]   1 4.103819e-05 0.0002856873
[3,]   2 6.225749e-05 0.0003633428
[4,]   3 9.293297e-05 0.0004589705
[5,]   4 1.365304e-04 0.0005758434
```

10.5.2   IRT Models

When using IRT models in the presmoothing step, **kequate** uses the Lord-Wingersky algorithm (described in Section 4.3.1) to obtain the estimated score probabilities. The previously created **mirt** objects ADMx.2PL and ADMy.2PL contain the IRT parameter estimates, which will be used in Section 10.7.2 as an input to the irtose() function to create the equating object m2pseSL. To obtain the estimated score probabilities, we can write similarly as we did for the log-linear models but using as argument the object m2pseSL, which contains the results of performing IRT KE, as follows:

```
> rj_irt <- getScores(m2pseSL)$X$r
> sk_irt <- getScores(m2pseSL)$Y$s
> cbind(0:80,rj_irt,sk_irt)
```

The resulting output of the estimated score probabilities for the first five score values, when the score probabilities have been estimated from a binary 2PL IRT model, appears as follows:

```
            rj_irt       sk_irt
[1,]   0 2.542969e-06 1.073803e-06
[2,]   1 1.650034e-05 7.682977e-06
[3,]   2 5.813646e-05 2.983366e-05
[4,]   3 1.483406e-04 8.367990e-05
[5,]   4 3.080168e-04 1.902467e-04
```

The marginal presmoothed score distributions of X and Y using the two different presmoothing methods as compared to the observed marginal score frequencies are shown in Figure 10.8. This figure was created using the code shown below, in which the first two lines use the margin() function from the **equate** package to obtain the marginal distributions from the previously created bivariate score frequency distributions feqX and feqY.

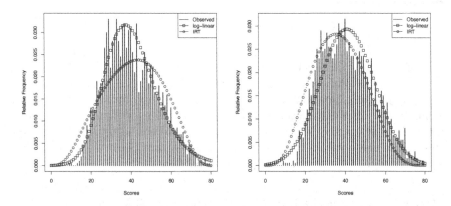

FIGURE 10.8: Presmoothing of marginal frequency distribution of X scores (left) and Y scores (right) for the ADM data using log-linear and IRT models.

```
> marX <- margin(feqX, margin=1)
> marY <- margin(feqY, margin=1)

> plot(0:80,as.matrix(marX)/sum(as.matrix(marX)),lwd=2.0,
+ xlab="Scores",ylab="Relative Frequency",type="h")
> points(0:80,rj,type="b",pch=0)
> points(0:80,rj_irt,type="b",pch=1)
> legend("topright",pch=c(NA,0,1),lty=c(1,1,1),
+ c("Observed","log-linear","IRT"))

> plot(0:80,as.matrix(marY)/sum(as.matrix(marY)),lwd=2.0,
+ xlab="Scores",ylab="Relative Frequency",type="h")
> points(0:80,sk,type="b",pch=0)
> points(0:80,sk_irt,type="b",pch=1)
> legend("topright",pch=c(NA,0,1),lty=c(1,1,1),
+ c("Observed","log-linear","IRT"))
```

It can be seen that both presmoothing methods lead to score distributions with shapes much *smoother* than those of the observed ones. The models, however, only managed to capture the main features of the data, but not all possible irregularities. Note that we could have used a more complicated log-linear model which could have captured the observed score distribution better.

10.6   Step 3: Continuization

We have seen in Section 5.3 that a continuized score can be defined as the sum of the original discrete score and a continuous random variable weighted by a bandwidth parameter, leading to the kernel that is to be used to obtain continuous approximations of the discrete score distributions. We can use a number of different kernels (Chapter 5) as well as a number of different methods to select the bandwidths (Chapter 6). When using **kequate**, these choices are made within the **kequate()** function that performs the equating. We will illustrate some of these options in Section 10.7.

10.6.1   Bandwidth Selection and Kernel Selection

In this chapter, we will illustrate the following different bandwidth selection methods: minimizing a penalty function (Section 6.3), cross-validation (Section 6.4), double smoothing (Section 6.6), and the rule-based bandwidth (Section 6.7). Further, we will also illustrate the following different kernels: Gaussian (Section 5.4.1), uniform (Section 5.4.3), and logistic (Section 5.4.2) kernels. Both the choice of kernel and the bandwidth selection method are directly fed into the **kequate()** function that is described in the next section.

10.7   Step 4: Equating

We have mentioned that both KE and IRT KE (i.e., KE when IRT models have been used in the presmoothing step) can be performed with the **kequate** package using the functions **kequate()** and **irtose()**, respectively. The former allows us to provide either user-specified score probabilities or a **glm** object where the fit of a log-linear model has been stored. The latter allows us to provide either the matrix of responses to each item or it can read either a **ltm** or a **mirt** object where the fit of an IRT model has been stored. In these functions, the user can decide which kernel continuization is to be used and how to select the bandwidths, as described in Chapters 5 and 6.

The general function call to **kequate()** is

```
kequate(design, ...)
```

where the argument **design** can take the following values: EG, SG, CB, NEAT_CE, NEAT_PSE, or NEC for the EG, SG, CB, CE, PSE, or NEC designs, respectively. Depending on the chosen design, different arguments can be used in **kequate()**, and for a list of all the currently implemented arguments, see Table 10.1, where all default values for the different arguments are given as well.

TABLE 10.1: List of arguments in `kequate()`.

Argument(s)	Designs	Description
x, y	ALL	Test score value vectors for test X and test Y.
a	NEAT_CE	Anchor test score value vector.
r, s	EG	Score probability vectors for tests X and Y. Or objects of class glm.
P	SG, NEAT_CE, NEAT_PSE, NEC	Matrix of bivariate score probabilities for tests X and Y (SG), tests X and A (NEAT CE, NEAT PSE), or test X and covariates (NEC) on population $\mathcal{P}$. Or an object of class glm.
Q	NEAT_CE, NEAT_PSE, NEC	Matrix of bivariate score probabilities for tests Y and A (NEAT CE, NEAT PSE) or test Y and covariates (NEC) on population $\mathcal{Q}$. Or an object of class glm.
P12, P21	CB	Matrices of bivariate score probabilities for tests X and Y. Or objects of class glm.
DMP, DMQ	NEAT_CE, NEAT_PSE, NEC	Design matrices for the used bivariate log-linear models on populations $\mathcal{P}$ and $\mathcal{Q}$, respectively (or groups taking test X and Y, respectively, in an EG design). Not needed if P and Q are of class glm.
DM	SG	Design matrix for the used bivariate log-linear model. Not needed if P is of class glm.
DM12, DM21	CB	Design matrices for the used bivariate log-linear models. Not needed if P12 and P21 are of class glm.
N	ALL	The sample size for population $\mathcal{P}$ (or the group taking test X in the EG design). Not needed if r, P, or P12 is of class glm.
M	EG, CB, NEAT_CE, NEAT_PSE, NEC	The sample size for population $\mathcal{P}$ (or the group taking test Y in the EG design). Not needed if s, Q, or P21 is of class glm.
w	NEAT_PSE	Optional argument to specify the weight given to population $\mathcal{P}$. Default is 0.5.
hx, hy, hxlin, hylin	EG, SG, CB, NEAT_PSE, NEC	Optional arguments to specify the continuization parameters manually.
hxP, hyQ, haP, haQ, hxPlin, hyQlin, haPlin, haQlin	NEAT_CE	Optional arguments to specify the continuization parameters manually.
wcb	CB	The weighting of the two test groups in the CB design. Default is 0.5.
KPEN	ALL	Optional argument to specify the constant in the penalty function. Default is 0.
wpen	ALL	Argument denoting at which point the derivatives in the second part of the penalty function should be evaluated. Default is 0.25.
linear	ALL	Logical argument indicating if only linear equating should be performed. Default is FALSE.
irtx, irty	ALL	Optional arguments to provide matrices of the probabilities of correctly answering the items on the parallel tests X and Y, as estimated in an IRT model.
smoothed	ALL	Logical argument denoting if the data provided are presmoothed or not. Default is TRUE.
kernel	ALL	A character vector denoting which kernel to use. Options are "gaussian," "logistic," "stdgaussian," and "uniform." Default is "gaussian."
slog	ALL	Logistic kernel parameter. Default is 1.
bunif	ALL	Uniform kernel parameter. Default is 0.5.
altopt	ALL	Logical argument to choose SRT bandwidth selection method. Default is FALSE.
DS	EG, SG, NEAT_CE, NEAT_PSE, NEC	Logical argument to choose DS bandwidth selection method. Default is FALSE.

The `irtose()` function works similarly to `kequate()` in various ways, but it also has unique arguments. This is why we present here a function call which includes more arguments with default values that can be used when performing IRT KE:

```
irtose(design="CE", P, Q, x, y, a=0, qpoints=seq(-6,6,by=0.1),
model="2pl", catsX=0, catsY=0, catsA=0, see="analytical",
replications=199, kernel="gaussian", h=list(hx=0, hy=0, hxP=0,
haP=0, hyQ=0, haQ=0), hlin=list(hxlin=0, hylin=0, hxPlin=0,
haPlin=0, hyQlin=0, haQlin=0), KPEN=0, wpen=0.5, linear=FALSE,
slog=1, bunif=1,altopt=FALSE, wS=0.5, eqcoef="mean-mean",
robust=FALSE,distribution = list("normal",
par = data.frame(mu = 0, sigma = 1)), DS = FALSE,CV = FALSE)}
```

Explanations of most of the different arguments are found in Table 10.1. This function call allows the user to read in objects from the **R** packages **ltm** or **mirt** using the arguments P and Q. These objects contain information about the estimated IRT model in population $\mathcal{P}$ (or $\mathcal{Q}$) or the response patterns in population $\mathcal{P}$ (or $\mathcal{Q}$). Quadrature points are used for calculating the marginal test score distribution, as specified in the argument `qpoints`. The default quadrature point values are a sequence from -6 to 6 with increment 0.1, as given in the above code. The argument `model` is used to specify the type of IRT model and the current options are the binary IRT models "2pl" (2PL) and "3pl" (3PL) and the polytomous models "GPCM" (GPCM) and "GRM" (GRM). The arguments `catsX`, `catsY`, and `catsA` are used with polytomous IRT models to define the number of response categories for each item in test forms X, Y and A, respectively.

Using the argument `see` we can specify whether to use bootstrap SEE or the default analytical SEE by writing either `see="bootstrap"` or `see="analytical"`. The analytical SEE is calculated using the delta method, as described in Section 8.1.4.1. The argument `replications` is used to set the number of bootstrap replications, and the default is 199. Under a NEAT design, we can specify which equating coefficients to use for the item parameter linking (see Section 7.6). The available options for `eqcoef` are `mean-mean`, `mean-gmean`, `mean-sigma`, `Haebara`, and `Stocking-Lord`.

The argument `robust` can be used for both the 2PL model and GPCM, and it is a logical indicator for whether a robust covariance matrix should be estimated. A robust covariance matrix refers to a covariance matrix that is robust against outliers. For more information about the different arguments, refer to Andersson et al. (2013) and the documentation of `kequate()` in CRAN.

10.7.1 Log-Linear Presmoothed Data

In the following example we first perform KE under the NEAT-PSE design when log-linear models have been used in the presmoothing step. We use the **glm** objects created in Section 10.4.1 and the default Gaussian kernel with the default penalty bandwidth selection method, where the weight of the synthetic population is set to the default value 0.5. In the code below, which executes the KE, we have specified the data collection design used, the possible scale score values on test forms X and Y, and the glm objects used.

```
> glmPSE <- kequate("NEAT_PSE",  0:80, 0:80, NEATvX, NEATvY)
```

The resulting summary output when excluding the score values $2, \ldots, 78$ looks as follows

```
> summary(glmPSE)
Design: NEAT/NEC PSE equipercentile

Kernel: gaussian

Sample Sizes:
  Test X: 2000
  Test Y: 2000

Score Ranges:
  Test X:
      Min = 0 Max = 80
  Test Y:
      Min = 0 Max = 80

Bandwidths Used:
        hx          hy      hxlin      hylin
1 0.6481502 0.6902837 12904.46 13329.32

Equating Function and Standard Errors:
  Score        eqYx       SEEYx
1      0 -0.9948293 0.1365356
2      1 -0.5296052 0.1782854
---
80    79 77.6029389 0.5194416
81    80 79.2837758 0.2746659

Comparing the Moments:
            PREYx
1  -0.0002357364
2  -0.0020577610
3  -0.0016381406
4   0.0016793712
5   0.0074088595
6   0.0145958021
7   0.0220455094
```

```
8    0.0284541804
9    0.0324977754
10   0.0328959256
```

The output yields information about which design has been used (NEAT PSE), that 2,000 test takers responded to each of the test forms, the scale score ranges (0:80 for both test forms), the bandwidths when the penalty method has been used (`hx`, `hy`), the bandwidths when the penalty method has been used if linear equating is performed (`hxlin`, `hylin`), the equated values (`eqYx`) with the SEE (`SEEYx`), and the PRE measure for the first ten moments (`PREYx`).

To perform equating using a CE transformation, we simply write as in the first line below, where the default Gaussian kernel and the penalty function method are used for the kernel and for selecting the bandwidth, respectively. If, however, we want to use a logistic kernel, or a uniform kernel with other bandwidth selection methods such as either double smoothing (see Section 6.6), cross-validation (see Section 6.4), or rule-based bandwidth selection (see Section 6.7), we can write as the lines below the first line.

```
> glmCE    <- kequate("NEAT_CE",0:80,0:80,0:40,NEATvX,NEATvY)
> glmCE1   <- kequate("NEAT_CE",0:80,0:80,0:40,NEATvX,NEATvY,
+ kernel="logistic")
> glmCEds <- kequate("NEAT_CE",0:80,0:80,0:40,NEATvX,NEATvY,
+ kernel="uniform", DS=1)
> glmCEcv <- kequate("NEAT_CE",0:80,0:80,0:40,NEATvX,NEATvY,
+ kernel="uniform", CV=1)
> glmCEsrt <- kequate("NEAT_CE",0:80,0:80,0:40,NEATvX,NEATvY,
+ kernel="uniform", altopt=TRUE)
```

The last argument in the function calls `DS=1`, `CV=1`, and `altopt=TRUE`, indicate that either double smoothing, cross-validation, or rule-based bandwidth selection has been used for selecting the bandwidths. As the summary outputs are similar, only the one for `glmCE` is shown below, where we have excluded the score values $2, \ldots, 78$.

```
Design: NEAT CE equipercentile

Kernel: gaussian

Sample Sizes:
    Test X: 2000
    Test Y: 2000

Score Ranges:
    Test X:
        Min = 0 Max = 80
```

```
Test Y:
    Min = 0 Max = 80
Test A:
    Min = 0 Max = 40
```

Bandwidths Used:
```
        hxP         hyQ        haP        haQ
1 0.6483522 0.6901055 0.4795392 0.4791372
```

```
 hxPlin   hyQlin   haPlin   haQlin
12896.49 13333.52 12368.8  12403.57
```

Equating Function and Standard Errors:
```
   Score        eqYx      SEEYx
1       0 -0.9949416 0.1363149
2       1 -0.5291419 0.1780668
---
80     79 77.6097805 0.5136455
81     80 79.2887844 0.2713237
```

Comparing the Moments:
```
            PREAx        PREYa
1   -0.013149815   0.07006645
2   -0.016587952  -0.12833451
3   -0.021565592  -0.37683680
4   -0.007367345  -0.75152590
5    0.022434448  -1.32311537
6    0.066843267  -2.14353841
7    0.125355036  -3.25047852
8    0.197695410  -4.66835537
9    0.283707926  -6.40798468
10   0.383303595  -8.46680454
```

The summary output yields information about which design has been used (NEAT CE), that 2,000 test takers answered each of the test forms, the scale score ranges (0:80 for test forms X and Y, and 0:40 for the anchor test form A), the bandwidths when the penalty method has been used (hxP, hyQ, haP, haQ), the bandwidths when the penalty method has been used if linear equating is performed (hxPlin, hyQlin, haPlin, haQlin), the equated values (eqYx) with the SEE (SEEYx), and the PRE measure for the first ten moments (PREAx, PREYa). Note that the largest difference in the summary output compared with the PSE summary output is that we get four different bandwidths and two sets of PRE values, i.e., the PRE is calculated both for the linking from X to A and the linking from A to Y.

10.7.2   IRT Presmoothed Data

To perform IRT KE under the NEAT-PSE design, we use the `irtose()` function in **kequate**. The function takes as inputs the design of interest, the IRT objects previously created by `mirt()` for the two test forms, the scale score values of the test forms and the anchor test form, which IRT model was used, and which method to use for the equating coefficients.

We start by exemplifying IRT KE with binary-scored response data. In the code shown below, we state which data collection design or method within the data collection design is used, which in this case is NEAT PSE (PSE). Further, we state that we use the previously obtained `mirt` IRT objects (ADMx.2PL, ADMy.2PL), the score scale range for test forms X, Y and A (0:80, 0:80, 0:40), which IRT model was used (2pl), and that we use the Stocking-Lord method for the equating coefficients (eqcoef). Note that for NEAT PSE with binary-scored items, we could also have used `mean-mean`, `mean-gmean`, `mean-sigma`, or `Haebara` for the equating coefficients, as described in Section 7.6.

```
> m2pseSL <- irtose("PSE", ADMx.2PL, ADMy.2PL, 0:80, 0:80,
+ 0:40, model= "2pl", eqcoef = "Stocking-Lord")
```

A summary of the output, when omitting results for score values $3, \ldots, 77$, reads as follows:

```
Design: IRT-OSE PSE

 Kernel: gaussian

 Sample Sizes:
   Test X: 2000
   Test Y: 2000

 Score Ranges:
   Test X:
       Min = 0 Max = 80
   Test Y:
       Min = 0 Max = 80

 Bandwidths Used:
           hx          hy
 1 0.6822623 0.6884574

 Equating Function and Standard Errors:
    Score      eqYx       SEEYx
 1      0   0.4030488 0.1656614
 2      1   1.4968474 0.2002586
 3      2   2.5699578 0.2277195
```

```
79    78 74.6731616 0.3114908
80    79 76.1552222 0.2784577
81    80 77.6813958 0.2380800
```

```
Comparing the Moments:
          PREYx
 1  -0.0001877879
 2  -0.0022512134
 3  -0.0088851601
 4  -0.0212147483
 5  -0.0398499551
 6  -0.0652838028
 7  -0.0979923341
 8  -0.1384419735
 9  -0.1870763929
10 -0.2443039997
```

The summary output shows similar information to when we have used kequate(), i.e., the design used, the sample sizes, the scale score ranges, the values of the bandwidth in the continuization step, the equated values with the corresponding SEE, and the PRE measures for the first 10 moments.

We can also perform IRT KE with a chained equating transformation irtose() by changing the first argument in the function to "CE", thus writing as follows:

```
> m2ce <- irtose("CE", ADMx.2PL, ADMy.2PL, 0:80, 0:80, 0:40)
```

A summary of the output, when omitting results for score values $3, \ldots, 77$, reads as follows:

```
Design: IRT-OSE CE

Kernel: gaussian

Sample Sizes:
   Test X: 2000
   Test Y: 2000

Score Ranges:
   Test X:
       Min = 0 Max = 80
   Test Y:
       Min = 0 Max = 80
```

```
Test A:
     Min = 0 Max = 40

Bandwidths Used:
          hxP         hyQ        haP        haQ
1 0.6851864 0.6853315 0.6533971 0.6399808


  hxPlin   hyQlin   haPlin   haQlin
13680.37 13511.37 6453.668 7251.469

Equating Function and Standard Errors:
    Score        eqYx       SEEYx
1        0  0.3559336 0.1728926
2        1  1.5116411 0.2119707
3        2  2.6378303 0.2395452
--
79      78 75.2013537 0.3262145
80      79 76.6113338 0.2957939
81      80 78.0870150 0.2563192


Comparing the Moments:
            PREAx          PREYa
1     4.117792e-05  0.0007874692
2    -1.196657e-03  0.0002457170
3    -3.604210e-03 -0.0086600218
4    -2.658304e-03 -0.0315341157
5     4.843506e-03 -0.0712225712
6     2.087383e-02 -0.1290658024
7     4.658746e-02 -0.2056422473
8     8.262863e-02 -0.3011402467
9     1.293266e-01 -0.4155448043
10    1.868172e-01 -0.5487350933
```

The summary output shows similar information to when we have used `kequate()` with NEAT CE, namely, the design used, the sample sizes, the score ranges, the values of the bandwidth in the continuization step, the equated values with the corresponding SEE, and the PRE measures for the first 10 moments.

Figures 10.9 and 10.10 show graphical comparisons of the KE and IRT KE transformations under the NEAT PSE design, and between those obtained under the NEAT CE using a Gaussian, a logistic, and a uniform kernel, respectively. From Figure 10.9 it can be seen that the equating transformations differ in most of the score scale. This result might be expected, because as was seen in Figure 10.8, the presmoothed score distributions using log-linear models and IRT models were in fact quite different. On the other hand, Figure 10.10 shows that the CE transformation that uses different kernel continuizations

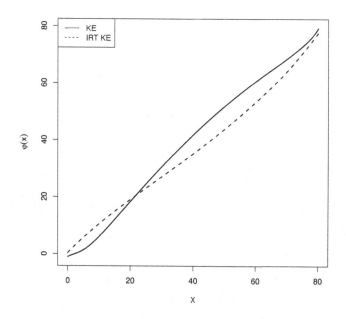

FIGURE 10.9: KE and IRT KE transformations under the NEAT PSE design.

are practically identical, indicating that the choice of kernel is not relevant in this case.

We will now illustrate how IRT KE with polytomous response data can be performed, using the previously simulated data and a GPCM. Recall that we had 40 unique items in each test form X and Y and 15 external anchor items in test form A, all of which were scored 0, 1, or 2. To conduct IRT KE with polytomously scored item data, we again use the irtose() function. In the code below, we state which data collection design or method within data collection design is used (CE), the previously created mirt objects (X55r, Y55r), the scale score ranges for test forms X, Y, and A (0:80,0:80,0:30), the vectors containing the number of possible response categories for test form X (catsX), test form Y (catsY), and the anchor test form (catsA), and which IRT model (GPCM) was used.

```
> CEP  <- irtose("CE", X55r,Y55r,0:80,0:80,0:30, catsX = rep(3,40),
+ catsY = rep(3,40), catsA = rep(3,15),model = "GPCM")
```

The summary of the output was excluded as its appearance is similar to the summary output from **kequate** when using binary-scored IRT models with the CE design.

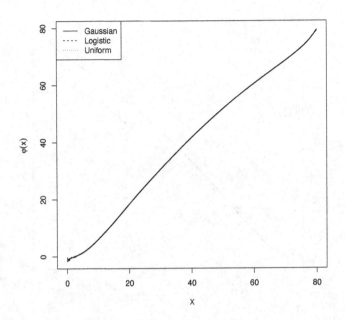

FIGURE 10.10: CE transformations for Gaussian, logistic, and uniform kernel continuization.

10.8   Step 5: Evaluating the Equating Transformation

We have several tools available for evaluating the equating transformation, as described in Chapter 8. These tools include both equating-specific measures and statistical measures. In this section, we will give examples of some of these tools.

10.8.1   PRE

The PRE, described in Section 8.1.2, measures the difference between the moments of the distribution for equated values with those of the score distribution to which scores are being equated, and it is thus favorable if the values are low. Note that the summary function yields the PRE in the last part of the output for a **kequate**-created equating object. The PRE can also be obtained from the previously created equating object **glmPSE** by using the getPre() function as follows.

```
> PREglmPSE <- getPre(glmPSE)
```

The PRE values for the first 10 moments can be seen in the last part of the output in Section 10.7.1, and are thus omitted here. All displayed PRE values were low, which means that the differences between the moments of the distribution of the equated values and those of the score distribution to which the scores are being equated, were small.

10.8.2 SEE

To obtain the SEE in **kequate**, we use the `getSee()` function together with the previously created equating object. The resulting output gives the SEE for each of the equated values. A graphical representation of the SEEs can be obtained by simply passing a `getSee()` object to the `plot()` function as shown in the code below. The resulting plot is shown in Figure 10.11.

```
> SeeADM <- getSee(glmCE)
> SeeADM
 [1]  0.1363149 0.1780668 0.2255348 0.2886961 0.3739439
 [6]  0.4765026 0.5782054 0.6658128 0.7337416 0.7808132
[11]  0.8087899 0.8209368 0.8207269 0.8111340 0.7948351
[16]  0.7743023 0.7511083 0.7260379 0.7000931 0.6743272
[21]  0.6491546 0.6247918 0.6016481 0.5800334 0.5602214
[26]  0.5421140 0.5257396 0.5114890 0.4992882 0.4889347
[31]  0.4803944 0.4735271 0.4682255 0.4643750 0.4618020
[36]  0.4603151 0.4597806 0.4600469 0.4608604 0.4622462
[41]  0.4641476 0.4663174 0.4689624 0.4717274 0.4748452
[46]  0.4781106 0.4815123 0.4852275 0.4891637 0.4933667
[51]  0.4979626 0.5030704 0.5088016 0.5152524 0.5225158
[56]  0.5308083 0.5403352 0.5509685 0.5630554 0.5762763
[61]  0.5908202 0.6062473 0.6229213 0.6406219 0.6590008
[66]  0.6779105 0.6969857 0.7158666 0.7342932 0.7518192
[71]  0.7677603 0.7813497 0.7917421 0.7977502 0.7975735
[76]  0.7885363 0.7665118 0.7245517 0.6496834 0.5136455
[81]  0.2713237
> plot(0:80, SeeADM, xlab="X scores", ylab ="SEE")
```

The SEE plot shown in Figure 10.11 has a similar appearance to others seen in, for example, González and Wiberg (2017) and von Davier et al. (2004) which were obtained for other data sets, i.e., the SEEs are larger at some of the lower and higher parts of the score scale, but smaller in the mid score range.

10.8.3 Bootstrap Standard Error of Equating

When performing IRT KE with the `irtose()` function, either analytical or bootstrap SEE can be obtained, the former being computed using the delta method by default. Below, we give the code for obtaining both analytical and

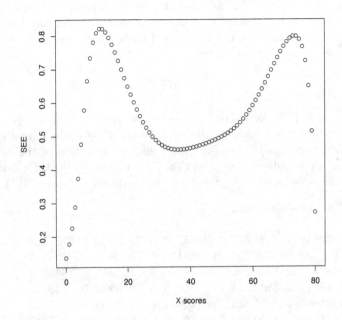

FIGURE 10.11: SEE for the NEAT CE.

bootstrap SEE when performing IRT KE under the NEAT design using a CE equating transformation.

In this example, we use the **ltm** package to fit the relevant IRT models. To obtain the SEE, we first fit 2PL models to the item response data in populations $\mathcal{P}$ and $\mathcal{Q}$ using the `ltm()` function and save these results in the objects `ADMxltm.2PL` and `ADMyltm.2PL`, respectively. The input for `ltm()` are binary data matrices with the answers to either the X or Y test forms in the first columns, and the answers to the anchor test form A in the last columns. The data matrix must be arranged in such a way that the unique test form data is in the first columns and the anchor test form data is in the following columns. This is needed to allow **kequate** to correctly read the **ltm** objects as inputs. The second and third lines in the the code below show how the data can be arranged correctly. The constructed data matrices are then read into `ltm()`, where we additionally indicate that the parameter estimates for the 2PL model are reported under the usual IRT parameterization by writing (`IRT.param=TRUE`) in the code below. For more details on the use of **ltm**, see Rizopoulos (2006), and the documentation available at CRAN.

The created **ltm** objects are then read into `irtose()` to conduct IRT KE with a CE transformation. Additionally, we need to indicate the score scale range for the different test forms, which IRT models should be used, and whether analytical or bootstrap SEE are to be computed. The default number of replications for the bootstrap SEE is 199, but we set it to 250 in the code below to illustrate that it is possible to choose a higher number of replications. It should be noted that running bootstrap SEE can take a considerable amount of time. Currently, bootstrap SEE is implemented in **kequate** when IRT models have been fitted using **ltm**, and when equating is performed under either the EG design or NEAT CE design.

```
> library(ltm)
> admP  <- as.matrix(ADMX[,c(41:120,1:40)])
> admQ  <- as.matrix(ADMY[,c(41:120,1:40)])
> ADMxltm.2PL <- ltm(admP~z1, IRT.param=TRUE)
> ADMyltm.2PL <- ltm(admQ~z1, IRT.param=TRUE)

> library(kequate)
> m2ceAN  <- irtose("CE", ADMxltm.2PL, ADMyltm.2PL, 0:80, 0:80,
+ 0:40, model= "2pl",  see="analytical")
> ANsee   <- getSee(m2ceAN)
> m2ceBoot<- irtose("CE", ADMxltm.2PL, ADMyltm.2PL, 0:80, 0:80,
+ 0:40, model= "2pl", see="bootstrap", replications=250)
> Bootsee <- getSee(m2ceBoot)
```

Figure 10.12 illustrates the difference between analytical and bootstrap SEE for our example. The SEEs are similar in the mid score range but the bootstrap SEEs are slightly larger in the lower and upper score ranges. The code to obtain this figure is given below.

```
> plot(ANsee,ylim=c(0,0.5),type='l',lty=1,
+ ylab="SEE",xlab="Scores")
> lines(Bootsee,col=4,lty=2)
> legend(box.lty=0,"topright",inset=0.02,
+ lty=c(1,2),col=c(1,4),c("Analytical","Bootstrap"))
```

10.8.4 SEED

The SEED measure (see Section 8.1.6.4) can be used to obtain the standard error for the differences between two KE transformations. A particular comparison of interest is made to decide whether a simpler linear equating transformation should be preferred to a more complicated KE transformation. We have seen in Section 7.4 that a KE transformation with a large bandwidth value approximates a linear equating transformation. The **kequate** package has different functions to perform these two type of comparison; genseed() and getSeed(), respectively. It is important to note that in the former case, the compared KE transformation functions must differ only in the bandwidth parameters that have been used in the estimation.

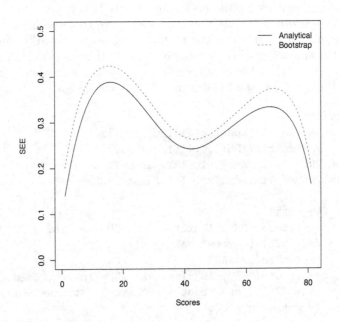

FIGURE 10.12: Analytical and bootstrap SEE comparison for IRT KE CE with the 2PL model.

In the example code below, we show how to obtain the SEED using `genseed()` when comparing the previously created KE transformations functions under the NEAT design; `glmPSE` and `glmCE`. In addition, using `getSeed()` we show how to obtain the SEED between the KE CE transformation and a linear version of it that uses a large value of the bandwidth.

```
> SeedADM       <- getSeed(glmCE)
> glmCEPSEseed <- genseed(glmCE,glmPSE)
> glmCEPSEseed
An object of class "genseed"
Slot "out":
          eqYxD        SEEDYx
1   -0.0001122601  0.007335755
2    0.0004632920  0.009860372
3    0.0011888379  0.013018772
---
79   0.0068175555  0.085129961
80   0.0068416522  0.063846269
81   0.0050086871  0.032290685
```

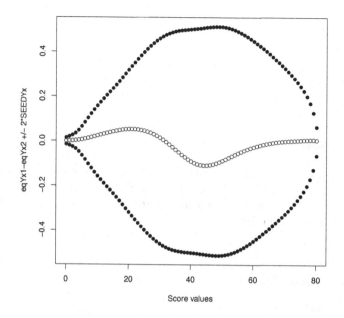

FIGURE 10.13: Comparison of the equated test scores of CE and PSE with a +/-2 SEED bound.

The resulting output gives the differences between the equated values from the two KE objects and the SEED where we have excluded results for rows 4-78. The SEED values ranged from 0.01 to 0.26, showing that the differences in equated values were in general small. An alternative graphical representation is shown in Figure 10.13, where the differences between the equated values have been plotted together with their uncertainty in terms of +/- 2SEED. This figure was obtained by simply writing plot(glmCEPSEseed). It can be seen that all differences are within the limits +/- 2SEED, indicating that there are no significant differences between the two compared KE transformation functions.

10.8.5 MSD, MAD, and RMSD

As described in Section 8.1.6, we can also use the MSD, MAD, and RMSD equating indices to compare two equating transformations computed differently. In the following example, we let $\hat{\varphi}_{Y1}(x_j)$ be the previously obtained KE function glmPSE, which we compare with $\hat{\varphi}_{Y2}(x_j)$ that represents the KE function stored in the object `glmCE`. The three indices were obtained using the following code, where `n` is the number of test score values.

```
> n    <- 81
> MSD  <- 1/n *sum(getEq(glmCE)-getEq(glmPSE))
> MAD  <- 1/n *sum(abs(getEq(glmCE)-getEq(glmPSE)))
> RMSD <- sqrt(1/n*sum(((getEq(glmCE)-getEq(glmPSE))^2)))
> MSD; MAD; RMSD
[1] -0.01374445
[1]  0.03768546
[1]  0.0502547
```

The outputs are single numbers, and the size of all these measures were low: -0.01 for MSD, 0.04 for MAD, and 0.05 for RMSD. This means that there were only small differences between the equated values obtained from the PSE and CE transformation functions.

10.9 Summary

In this chapter, we have given examples of how to implement the GKE framework under the NEAT design with different alternatives in each of the five steps. Throughout the chapter, we have used binary-scored items from a real college admissions test and simulated data for polytomously scored items. The GKE steps were conducted in the **R** package **kequate**, where several options to conduct GKE are included.

Most of the illustrated options are also available for the EG design. Although we have demonstrated several possible choices in GKE, it should be emphasized that the selected options depend on different situations and which assumptions can be made about the data. Although we have provided some examples here, they are by no means exhaustive, and many other combinations of the different alternatives described in the book could of course have been used.

Part IV

Appendix

A

Installing **R** and Reading in Data

In this appendix, we briefly describe how to install the free software environment **R** (R Core Team, 2024) and how to read in data files with item-level test data.

A.1 Installing R for the First Time

R can be used with Windows, MacOS, and Unix platforms and can be downloaded from the Comprehensive R Archive Network (CRAN) webpage http://cran.r-project.org/ or preferably from one of the mirror sites available that are near the place where the software is being installed. Precompiled binary distributions of the base package and contributed packages are available for different platforms.

The examples used in this book have been tested on both MacOS and Windows platforms using **R** version 4.3.3. To install **R** for the first time, the executable file for either Windows (R-4.3.3-win.exe) or for MacOS (R-4.3.3-arm64.pkg) should be downloaded from CRAN. Once downloaded, double clicking on the file will start the setup wizard to install **R**.

A.1.1 Rstudio

Rstudio is a graphical user interface that can be downloaded from http://www.rstudio.org/, and which should be installed after the **R** software has previously been installed. Besides providing the usual **R** console, **Rstudio** has several menus with more alternatives, enabling, for example, a history and a workspace window. **Rstudio** also facilitates access to files and packages and has a plot tab that facilitates the export of plots to other software.

A.2   Installing and Using R packages

Once **R** is installed, contributed **R** packages can be added to the system. To install a package, one can go to the tab "Packages/Install package(s)", choose the package of interest (e.g. **kequate**), and press ok to install the package.

To be able to use the functions within the installed package, it first needs to be loaded by typing `library(kequate)` in the command prompt, or going to the menu and selecting "Packages/load package". If help is needed with a function, one can write `help()` in the **R** command prompt and include within the parenthesis the name of the function for which help is needed. For general help regarding **R**, there are many online resources as well as books that are useful for learning **R**. If specific help with **R** in the context of equating is needed, González and Wiberg (2017) is a relevant source.

The **R** packages used in the examples in this book have been executed under the following versions:

- **kequate** version 1.6.4

- **SNSequate** version 1.3-5

- **ltm** version 1.2.0

- **mirt** version 1.41

- **psych** version 2.4.1

- **equate** version 2.0.8

A.3   Loading Data

The typical data set to be used is assumed to have a structure in which each row represents a test taker and each column represents an item. To load a data set saved as ADM.Rda in the working directory, the `load()` function can be used as shown in the first code line below. Data stored with different file formats can also be loaded by using the functions `read.table()` or `read.csv()` as follows:

```
> load(ADM.Rda)
> ADMt <- read.table(file="ADMt.txt",sep="\t",header=TRUE)
> ADMd <- read.table("ADMd.dat",header=T)
> ADMc <- read.csv("ADMc.csv",sep=";",header=T)
```

The second line in this code illustrates how to read in a .txt file that is separated with commas and has headers. Similarly, the third and fourth lines can be used when reading in .dat or .csv files. Data can also be read directly from an Excel file (.xls) if the **R** package **XLConnect** is installed, by using the loadWorkbook() and readWorksheet() functions. We can also read in an SPSS file (.sav) using the **R** package **foreign** and the read.spss() function.

Finally, we recommend visualizing the data set (for instance, the ADM data) to assure that it has been read in correctly, by writing View(ADM).

B

R packages for GKE

To perform GKE, we recommend using **R** and a number of useful packages. Note that there are also other software programs that can be used to implement some steps of the GKE framework. For example, the KE software (ETS, 2004) covers the original KE method, and *equating recipes* (Brennan et al., 2009) also implements the original KE method, as well as presmoothing with beta4 models among other things. Here, we briefly describe **R** packages that are currently available to perform the different methods in the GKE framework. The **R** packages discussed are **kequate** (Andersson et al., 2013), **SNSequate** (González, 2014), **equate** (Albano, 2016), **ltm** (Rizopoulos, 2006), and **mirt** (Chalmers, 2012). Note that we are not giving examples using these packages here, but we only point out which GKE methods are currently available with the different packages.

B.1 Presmoothing

Presmoothing in GKE can be done using log-linear models, IRT models, beta4 models, and discrete kernels.

To presmooth data with log-linear models, there are several options. For instance, the general `glm` function in **R** can be used to fit log-linear models. The obtained `glm` objects can then be read into **kequate**. This is illustrated in Section 10.4.1 and other examples are given in Section 4.2.2 in González and Wiberg (2017). It is also possible to use the `loglin.smooth()` function in **SNSequate**, as exemplified in Section 4.2.1 in González and Wiberg (2017). A third possibility would be to use the `presmoothing()` function in the **equate** package, and an example can be found in Chapter 2 in González and Wiberg (2017).

To presmooth the score data with IRT models, one can use either **ltm** or **mirt**. The resulting objects can be read into **kequate**. Examples with **mirt**, using binary scored items, are given in Section 10.4.2, and with polytomously scored items in Section 10.4.3. Examples with **ltm** can be found in González and Wiberg (2017); for binary scored items, refer to Section 7.3.3 and for polytomously scored items, refer to Section 7.3.4 in that book.

SNSequate can be used to presmooth the score data with beta4 models and an example is given in Section 9.3.1. Finally, **SNSequate** can also be used to presmooth the score data with discrete kernel estimators, and an example is given in Section 9.3.2.

B.2 Estimating Score Probabilities

The estimation of score probabilities can be done with either **kequate** or **SNSequate**, but it depends on the choice of the presmoothing model. Examples using **SNSequate** under the EG design for the beta4 model can be found in Section 9.4.1 and for discrete kernel estimators in Section 9.4.2. Examples using **kequate** under the NEAT design can be found in Section 10.5.1 for log-linear models and for IRT models in Section 10.5.2. Other examples of estimation of score probabilities using **kequate** and **SNSequate** can be found in González and Wiberg (2017) in Sections 4.3.2 and 4.3.1, respectively.

B.3 Continuization

Either **kequate** or **SNSequate** can be used for implementing a Gaussian, logistic, or uniform kernel continuization. Examples using **kequate** are given in Section 10.7.1. **SNSequate** has to be used if either adaptive kernels or the Epanechnikov kernel are used, and examples of those are given in Section 9.5.1.

B.4 Bandwidth Selection

Several different methods can be used to select a bandwidth. The most common method is to minimize a penalty function, and this method is available in both **kequate** and **SNSequate**. Examples when minimizing a penalty function with **kequate** can be found in Section 10.7.1 and when using **SNSequate** in Section 9.5.1. The bandwidth selection methods, leave-one-out cross-validation, double smoothing, and rule-based bandwidth selection, can only be found in **kequate**, and examples using these methods are given in Section 10.7.1.

B.5 Equating

To perform GKE with the EG, SG, CB, or NEAT designs, we can use either **kequate** or **SNSequate**. If log-linear models have been chosen in the presmoothing step, both packages can be used for the EG, SG, CB, and NEAT designs. An example of performing equating when log-linear models have been used in the presmoothing step for the NEAT design is given for **kequate** in Chapter 10. Other examples using the EG, SG, NEAT, and NEC designs when log-linear models have been used in the presmoothing using **kequate** are given in Section 4.5 2 in González and Wiberg (2017). Examples with the EG, and CB designs when log-linear models have been used in the presmoothing step using **SNSequate** are given in Section 4.5.1 in González and Wiberg (2017).

When beta4 models or discrete kernels have been chosen in the presmoothing step, only **SNSequate** can be used. Examples of performing equating when beta4 or discrete kernels have been chosen as presmoothing models with the EG design using **SNSequate** are given in Chapter 9.

If IRT models have been chosen in the presmoothing step, only **kequate** can be used and an example with the NEAT design is given in Chapter 10.

Local equating within GKE can be done with **kequate** and González and Wiberg (2017) provides an example of the use of local observed-score KE in Section 6.4.4.2, and the use of local IRT observed-score KE in Section 6.4.4.3.

B.6 Evaluating the Equating Transformation

To evaluate the equating transformation, we can use either equating-specific measures or statistical measures, or both.

The equating-specific measure DTM and the equating indices can easily be programmed directly in **R**. The equating-specific measures SEE, SEED, and PRE are available as separate functions in both **kequate** and **SNSequate**.

The statistical measures (bias, SE, MSE, and RMSE) can easily be programmed within **R**, and either **kequate** or **SNSequate** can be used to estimate the equating transformations needed in these measures. We recommend using the package that was used in the equating step.

Bibliography

Adroher, N. D., S. Kreiner, C. Young, R. Mills, and A. Tennant (2019). Test equating sleep scales: Applying the Leunbach's model. *BMC Medical Research Methodology 19*, 1–13.

Akaike, H. (1981). Likelihood of a model and information criteria. *Journal of Econometrics 16*(1), 3–14.

Albano, A. D. (2016). equate: An R package for observed-score linking and equating. *Journal of Statistical Software 74*(8), 1–36.

Albano, A. D. and M. Wiberg (2019). Linking with external covariates: Examining accuracy by anchor type, test length, ability difference, and sample size. *Applied Psychological Measurement 43*(8), 597–610.

Andersson, B. (2017). Asymptotic standard errors of observed-score equating with polytomous IRT models. *Journal of Educational Measurement 53*(4), 459–477.

Andersson, B., K. Bränberg, and M. Wiberg (2013). Performing the kernel method of test equating with the package kequate. *Journal of Statistical Software 55*(6), 1–25.

Andersson, B. and A. A. von Davier (2014). Improving the bandwidth selection in kernel equating. *Journal of Educational Measurement 51*(3), 223–238.

Andersson, B. and M. Wiberg (2014). IRT observed-score kernel equating with the R package kequate. *R Vignette: https://cran.r-project.org/web/packages/kequate/vignettes/irtguide.pdf*, 1–12.

Andersson, B. and M. Wiberg (2017). Item response theory observed-score kernel equating. *Psychometrika 82*(1), 48–66.

Angoff, W. H. (1971). Scales, norms and equivalent scores. In R. L. Thorndike (Ed.), *Educational Measurement (2nd ed.)*, pp. 508–600. Washington, DC: American Council on Education. (Reprinted as Angoff W. H. (1984). *Scales, Norms and Equivalent Scores*. Princeton, NJ: Educational Testing Service.).

Bahadur, R. R. (1966). A note on quantiles in large samples. *The Annals of Mathematical Statistics 37*(3), 577–580.

Baker, F. and S. Kim (2004). *Item Response Theory: Parameter Estimation Techniques*. New York: Marcel Dekker.

Barrett, M. D. and W. J. van der Linden (2019). Estimating linking functions for response model parameters. *Journal of Educational and Behavioral Statistics 44*(2), 180–209.

Beyer, W. (1982). *CRC Standard Mathematical Tables*, Volume 28. Boca Raton, FL: CRC Press.

Birnbaum, A. (1968). Some latent trait models and their use in inferring any examinee's ability. In F. M. Lord and M. R. Novick (Eds.), *Statistical Theories of Mental Test Scores*, pp. 395–479. Reading, MA: Adison-Wesley.

Bishop, Y., S. Fienberg, and P. Holland (1975). *Discrete Multivariate Analysis: Theory and Practice*. Cambridge, MA: MIT Press.

Bock, R. and M. Aitkin (1981). Marginal maximum likelihood estimation of item parameters: Application of an EM algorithm. *Psychometrika 46*, 443–459.

Braun, H. and P. Holland (1982). Observed-score test equating: a mathematical analysis of some ETS equating procedures. In P. Holland and D. Rubin (Eds.), *Test Equating*, Volume 1, pp. 9–49. New York: Academic Press.

Brennan, M., T. Wang, S. Kim, and T. Seol (2009). Equating recipes. Technical report, CASMA Research Report.

Brennan, R. L. and M. J. Kolen (1987). Some practical issues in equating. *Applied Psychological Measurement 11*(3), 279–290.

Casella, G. and R. L. Berger (2002). *Statistical Inference*, Volume 2. Pacific Grove, CA: Duxbury.

Cattell, R. B. (1966). The scree test for the number of factors. *Multivariate Behavioral Research 1*(2), 245–276.

Chalmers, R. P. (2012). mirt: A multidimensional item response theory package for the R environment. *Journal of Statistical Software 48*(6), 1–29.

Charalambides, C. A. (2005). *Combinatorial Methods in Discrete Distributions*. John Wiley & Sons: Hoboken, NJ.

Chen, H. (2012). A comparison between linear IRT observed-score equating and Levine observed-score equating under the generalized kernel equating framework. *Journal of Educational Measurement 49*, 269–284.

Chen, W.-H. and D. Thissen (1997). Local dependence indexes for item pairs using item response theory. *Journal of Educational and Behavioral Statistics 22*(3), 265–289.

Chiu, S.-T. (1996). A comparative review of bandwidth selection for kernel density estimation. *Statistica Sinica*, 129–145.

Christensen, K. B., J. B. Bjorner, S. Kreiner, and J. H. Petersen (2002). Testing unidimensionality in polytomous Rasch models. *Psychometrika 67*(4), 563–574.

Cid, J. A. and A. A. von Davier (2015). Examining potential boundary bias effects in kernel smoothing on equating: an introduction for the adaptive and Epanechnikov kernels. *Applied Psychological Measurement 39*(3), 208–222.

Cook, L. L. and D. Eignor (1991). IRT equating methods. *Educational Measurement: Issues and Practice 10*(3), 37–45.

Cox, D. R. and D. V. Hinkley (1974). *Theoretical Statistics*. London, UK: Chapman and Hall.

Cui, Z. and M. J. Kolen (2009). Evaluation of two new smoothing methods in equating: The cubic B-spline presmoothing method and the direct presmoothing method. *Journal of Educational Measurement 46*(2), 135–158.

Darroch, J. N. (1964). On the distribution of the number of successes in independent trials. *The Annals of Mathematical Statistics 35*, 1317–1321.

De Boeck, P. and M. Wilson (2004). *Explanatory Item Response Models: A Generalized Linear and Nonlinear Approach*. New York: Springer.

del Roo, A. Q. and G. Estevez-Perez (2012). Nonparametric kernel distribution function estimation with kerdiest: An R package for bandwidth choice and applications. *Journal of Statistical Software 50*(8), 1–21.

Denuit, M. and P. Lambert (2005). Constraints on concordance measures in bivariate discrete data. *Journal of Multivariate Analysis 93*(1), 40–57.

Doksum, K. (1974). Empirical probability plots and statistical inference for nonlinear models in the two-sample case. *The Annals of Statistics 2*, 267–277.

Dorans, N. and M. Feigenbaum (1994). Equating issues engendered by changes to the SAT and PSAT/NMSQT. *Technical Issues Related to the Introduction of the New SAT and PSAT/NMSQT 1994*(10), 91–122.

Dorans, N. and P. Holland (2000). Population invariance and the equatability of tests: Basic theory and the linear case. *Journal of Educational Measurement 37*(4), 281–306.

Dorans, N. J., M. Pommerich, and P. W. Holland (2007). *Linking and Aligning Scores and Scales*. New York: Springer.

Drasgow, F., M. V. Levine, and E. A. Williams (1985). Appropriateness measurement with polychotomous item response models and standardized indices. *British Journal of Mathematical and Statistical Psychology 38*(1), 67–86.

Du, K.-L., M. Swamy, et al. (2016). *Search and Optimization by Metaheuristics.* Springer: Birkhäuser Cham.

Duong, M. and A. von Davier (2013). Heterogeneous populations and multistage test design. In R. Millsap, L. van der Ark, D. Bolt, and C. Woods (Eds.), *New Developments in Quantitative Psychology. Springer Proceedings in Mathematics and Statistics*, Volume 66, pp. 275–286. New York: Springer.

Efron, B. (1982). *The Jackknife, the Bootstrap and Other Resampling Plans*, Volume 38. Philadelphia: SIAM.

Efron, B. and R. Tibshirani (1993). *An Introduction to the Bootstrap.* London: Chapman & Hall.

Embrechts, P. and M. Hofert (2013). A note on generalized inverses. *Mathematical Methods of Operations Research 77*(3), 423–432.

Epanechnikov, V. (1969). Non-parametric estimation of a multivariate probability density. *Theory of Probability and its Applications 14*, 153–158.

ETS (2004). KE software. Technical report, Princeton.

Everitt, B. (1984). *An Introduction to Latent Variable Models.* London: Chapman and Hall.

Fairbank, B. (1987). The use of presmoothing and postsmoothing to increase the precision of equipercentile equating. *Applied Psychological Measurement 11*(3), 245–262.

Fernandez, M. and S. Williams (2010). Closed-form expression for the Poisson-Binomial probability density function. *IEEE Transactions on Aerospace and Electronic Systems 46*, 803–816.

Fischer, G. and I. Molenaar (1995). *Rasch Models: Foundations and Recent Developments.* New York: Springer.

Fisher, R. A. (1922). On the mathematical foundations of theoretical statistics. *Philosophical Transactions of the Royal Society of London, Series A 222*, 309–368.

Fisher, R. A. and J. Wishart (1932). The derivation of the pattern formulae of two-way partitions from those of simpler patterns. *Proceedings of the London Mathematical Society s2-33*(1), 195–208.

Glas, C. A. and N. D. Verhelst (1995). Tests of fit for polytomous Rasch models. In G. Fischer and I. Molenaar (Eds.), *Rasch Models: Foundations, Recent Developments, and Applications*, pp. 325–352. Berlin: Springer-Verlag.

González, J. (2014). SNSequate: Standard and nonstandard statistical models and methods for test equating. *Journal of Statistical Software 59*(7), 1–30.

González, J. and R. Gempp (2021). *Test Comparability and Measurement Validity in Educational Assessment*, pp. 173–204. Cham: Springer International Publishing.

González, J. and E. San Martín (2018). An alternative view on the NEAT design in test equating. In M. Wiberg, S. Culpepper, R. Janssen, J. González, and D. Molenaar (Eds.), *Quantitative Psychology. The 82nd Annual Meeting of the Psychometric Society, Zurich, Switzerland, 2017*, pp. 111–120. Springer.

González, J. and E. San Martín (2024). Linear statistical test equating models under the NEAT design. *Manuscript submitted for publication*.

González, J., E. San Martín, and I. Varas (2024). On the property of symmetry of the equipercentile transformation function in test equating. *Manuscript submitted for publication*.

González, J. and A. A. von Davier (2017). An illustration of the Epanechnikov and adaptive continuization methods in kernel equating. In L. van der Ark, M. Wiberg, S. Culpepper, J. Douglas, and W.-C. Wang (Eds.), *Quantitative Psychology. The 81st Annual Meeting of the Psychometric Society, Asheville, North Carolina, 2016*, pp. 253–262. Springer.

González, J. and M. von Davier (2013). Statistical models and inference for the true equating transformation in the context of local equating. *Journal of Educational Measurement 50*(3), 315–320.

González, J. and G. Wallin (2021). An illustration on the quantile-based calculation of the standard error of equating in kernel equating. In M. Wiberg, D. Molenaar, J. González, U. Böckenholt, and J. Kim (Eds.), *Quantitative Psychology: The 85th Annual Meeting of the Psychometric Society, Virtual*, pp. 233–241. Springer.

González, J. and M. Wiberg (2017). *Applying Test Equating Methods using R*. Cham: Springer.

González, J. and M. Wiberg (2024a). A familiy of discrete kernels for presmoothing test score distributions. In M. Wiberg, J. Kim, H. Hwang, H. Wu, and T. Sweet (Eds.), *Quantitative Psychology: The 88th Annual Meeting of the Psychometric Society, Maryland, USA, 2023*, pp. 1–13. Springer.

González, J. and M. Wiberg (2024b). The performance of different mixture models for presmoothing score distributions in kernel equating. *Manuscript submitted for publication*.

González, J. and M. Wiberg (2024c). Presmoothing score distributions using discrete kernels. *Manuscript submitted for publication*.

González, J., M. Wiberg, and A. A. von Davier (2016). A note on the Poisson's binomial distribution in item response theory. *Applied Psychological Measurement 40*(4), 302–310.

Gross, A. L., A. M. Kueider-Paisley, C. Sullivan, D. Schretlen, and I. N. N. D. Initiative (2019). Comparison of approaches for equating different versions of the mini-mental state examination administered in 22 studies. *American Journal of Epidemiology 188*(12), 2202–2212.

Haberman, S. (2011). Using exponential families for equating. In A. von Davier (Ed.), *Statistical Models for Test Equating, Scaling, and Linking*, pp. 125–140. New York: Springer.

Haberman, S. J. (2015). Pseudo-equivalent groups and linking. *Journal of Educational and Behavioral Statistics 40*(3), 254–273.

Haebara, T. (1980). Equating logistic ability scales by a weighted least squares method. *Japanese Psychological Research 22*, 144–149.

Häggström, J. and M. Wiberg (2014). Optimal bandwidth selection in observed-score kernel equating. *Journal of Educational Measurement 51*(2), 201–211.

Hall, P. (2001). Biometrika centenary: Nonparametrics. *Biometrika 88*(1), 143–165.

Hall, P., J. Marron, and B. U. Park (1992). Smoothed cross-validation. *Probability Theory and Related Fields 92*(1), 1–20.

Hambleton, R. K. and H. Swaminathan (1985). *Item Response Theory: Principles and Applications*. Dordrecht: Kluwer Nijhoff Publishing.

Han, T., M. Kolen, and J. Pohlmann (1997). A comparison among IRT true- and observed-score equatings and traditional equipercentile equating. *Applied Measurement in Education 10*(2), 105–121.

Handing, E., S. R. Rapp, S.-H. Chen, J. Rejeski, M. Wiberg, K. Bandeen-Roche, S. Craft, D. Kitzman, and E. H. Ip (2021). Heterogeneity in association between cognitive function and gait speed among older adults: An integrative data analysis study. *The Journals of Gerontology: Series A 76*(4), 710–715.

Hanson, B. (1991a). Method of moments estimates for the four-parameter beta compound binomial model and the calculation of classification consistency indexes. *ACT Research Report 1991*(5), i–21.

Hanson, B. (1996). Testing for differences in test score distributions using loglinear models. *Applied Measurement in Education 9*(4), 305–321.

Hanson, B. and R. L. Brennan (1990). An investigation of classification consistency indexes estimated under alternative strong true score models. *Journal of Educational Measurement 27*(4), 345–359.

Hanson, B., L. Zeng, and D. Colton (1994). A comparison of presmoothing and postsmoothing methods in equipercentile equating. *ACT Research Report 1994*(4), i–36.

Hanson, B. A. (1991b). A comparison of bivariate smoothing methods in common-item equipercentile equating. *Applied Psychological Measurement 15*(4), 391–408.

Härdle, W. (1991). *Smoothing Techniques with Implementation in S*. New York, NY: Springer.

Harris, D. J. and J. D. Crouse (1993). A study of criteria used in equating. *Applied Measurement in Education 6*(3), 195–240.

Hart, J. D. (1997). *Nonparametric Smoothing and Lack-of-Fit Tests*. New York: Springer.

Heidenreich, N.-B., A. Schindler, and S. Sperlich (2013). Bandwidth selection for kernel density estimation: A review of fully automatically selectors. *Advances in Statistical Analysis 97*(4), 403–433.

Hoeffding, W. (1956). On the distribution of the number of successes in independent trials. *The Annals of Mathematical. Statistics 27*, 713–721.

Holland, P. and N. Dorans (2006). Linking and equating. In R. L. Brennan (Ed.), *Educational Measurement (4th ed.)*, pp. 187–220. Westport, CT: Praeger.

Holland, P., B. King, and D. Thayer (1989). The standard error of equating for the kernel method. Technical Report 89-83, Educational Testing Service.

Holland, P. and D. Rubin (1982). *Test Equating*. New York: Academic Press.

Holland, P. and D. Thayer (1989). The kernel method of equating score distributions. Technical report, Princeton, NJ: Educational Testing Service.

Holland, P. and D. Thayer (2000). Univariate and bivariate loglinear models for discrete test score distributions. *Journal of Educational and Behavioral Statistics 25*(2), 133–183.

Holland, P. W., N. J. Dorans, and N. S. Petersen (2006). Equating test scores. In C. Rao and S. Sinharay (Eds.), *Psychometrics*, Volume 26 of *Handbook of Statistics*, pp. 169 – 203. Elsevier.

Holland, P. W. and D. T. Thayer (1987). Notes on the use of log-linear models for fitting discrete probability distributions. *ETS Research Report Series 1987*(2), i–40.

Hong, Y. (2013). On computing the distribution function for the Poisson binomial distribution. *Computational Statistics and Data Analysis 59*, 41–51.

Hooke, R. and T. A. Jeeves (1961). " Direct search" solution of numerical and statistical problems. *Journal of the ACM 8*(2), 212–229.

Horn, J. L. (1965). A rationale and test for the number of factors in factor analysis. *Psychometrika 30*(2), 179–185.

Hu, L.-T. and P. M. Bentler (1999). Cutoff criteria for fit indexes in covariance structure analysis: Conventional criteria versus new alternatives. *Structural Equation Modeling: A Multidisciplinary Journal 6*(1), 1–55.

Jiang, Y., A. A. von Davier, and H. Chen (2012). Evaluating equating results: Percent relative error for chained kernel equating. *Journal of Educational Measurement 49(1)*, 39–58.

Jonas, M., J. Marron, and S. Sheather (1996). A brief survey of bandwidth selection for density estimation. *Journal of the American Statistical Association 91*(433), 401–407.

Kang, H.-A., Y.-H. Su, and H.-H. Chang (2018). A note on monotonicity of item response functions for ordered polytomous item response theory models. *British Journal of Mathematical and Statistical Psychology 71*(3), 523–535.

Keats, J. A. and F. M. Lord (1962). A theoretical distribution for mental test scores. *Psychometrika 27*(1), 59–72.

Keller, L. A. and R. K. Hambleton (2013). The long-term sustainability of irt scaling methods in mixed-format tests. *Journal of Educational Measurement 50*(4), 390–407.

Kendall, M. and A. Stuart (1997). *The Advanced Theory of Statistics* (4th ed.), Volume 1. London: Griffin.

Kim, S., D. J. Harris, and M. J. Kolen (2010). Equating with polytomous item response models. In M. L. Nering and R. Ostini (Eds.), *Handbook of Polytomous Item Response Theory Models*, Volume 1, pp. 257–291. New York: Routledge.

Kirk, R. E. (1996). Estimating linking functions for response model parameters. *Educational and Psychological Measurement 56*(5), 746–759.

Köhler, M., A. Schindler, and S. Sperlich (2014). A review and comparison of bandwidth selection methods for kernel regression. *International Statistical Review 82*(2), 243–274.

Kokonendji, C. C. and T. S. Kiessé (2011). Discrete associated kernels method and extensions. *Statistical Methodology 8*(6), 497 – 516.

Kokonendji, C. C., T. Senga Kiesse, and S. S. Zocchi (2007). Discrete triangular distributions and non-parametric estimation for probability mass function. *Journal of Nonparametric Statistics 19*(6-8), 241–254.

Kolen, M. (1984). Effectiveness of analytic smoothing in equipercentile equating. *Journal of Educational and Behavioral Statistics 9*(1), 25–44.

Kolen, M. (1991). Smoothing methods for estimating test score distributions. *Journal of Educational Measurement 28*(3), 257–282.

Kolen, M. and R. Brennan (2014). *Test Equating, Scaling, and Linking: Methods and Practices* (3rd ed.). New York: Springer.

Kolen, M. and D. Jarjoura (1987). Analytic smoothing for equipercentile equating under the common item nonequivalent populations design. *Psychometrika 52*, 43–59.

Kolen, M. J., B. A. Hanson, and R. L. Brennan (1992). Conditional standard errors of measurement for scale scores. *Journal of Educational Measurement 29*(4), 285–307.

Kolen, M. J., L. Zeng, and B. Hanson (1996). Conditional standard errors of measurement for scale scores using IRT. *Journal of Educational Measurement 33*(2), 129–140.

Laukaityte, I. and M. Wiberg (2022). How to choose the anchor test when equating test scores. *Paper presented at AEA-Europe, Dublin 9-12 November, 2022*, 1–10.

Laukaityte, I. and M. Wiberg (2024). The impact of differences in group abilities and anchor test features on three non-IRT test equating methods. *Practical Assessment, Research, and Evaluation 29*(5), 1–23.

Le Cam, L. (1960). An approximation theorem for the Poisson binomial distribution. *Pacific Journal of Mathematics 10*, 1181–1197.

Lee, E., W. Lee, and R. Brennan (2010). Assessing equating results based on first-order and second-order equity. Technical report, CASMA Research Report.

Lee, Y. and A. von Davier (2011). Equating through alternative kernels. In A. von Davier (Ed.), *Statistical Models for Test Equating, Scaling, and Linking*, Volume 1, pp. 159–173. New York: Springer.

Lee, Y.-H. and A. A. von Davier (2013). Monitoring scale scores over time via quality control charts, model-based approaches, and time series techniques. *Psychometrika 78*(3), 557–575.

Lehmann, E. L. (1999). *Elements of Large-Sample Theory*. New York: Springer.

Leoncio, W. and M. Wiberg (2018). Evaluating equating transformations from different frameworks. In M. Wiberg, S. Culpepper, R. Janssen, J. González, and D. Molenaar (Eds.), *Quantitative Psychology. The 82nd Annual Meeting of the Psychometric Society, Zurich, Switzerland, 2017*, pp. 101–110. Springer.

Leoncio, W., M. Wiberg, and M. Battauz (2023). Evaluating equating transformations in IRT observed-score and kernel equating methods. *Applied Psychological Measurement 47*(2), 123–140.

Li, D., Y. Jiang, and A. A. von Davier (2012). The accuracy and consistency of a series of IRT true score equatings. *Journal of Educational Measurement 49*(2), 167–189.

Liang, T. and A. A. von Davier (2014). Cross-validation: An alternative bandwidth-selection method in kernel equating. *Applied Psychological Measurement 38*(4), 281–295.

Lindsay, B. G. (1995). *Mixture Models: Theory, Geometry, and Applications*. Institute for Mathematical Statistics: Hayward, CA.

Liou, M., P. E. Cheng, and E. G. Johnson (1997). Standard error of the kernel equating methods under the common-item design. *Applied Psychological Measurement 21*(4), 349–369.

Little, R. and D. Rubin (1994). Test equating from biased samples, with application to the Armed Services Vocational Aptitude Battery. *Journal of Educational and Behavioral Statistics 19*(4), 309–335.

Liu, C. and M. J. Kolen (2011). Evaluating smoothing in equipercentile equating using fixed smoothing parameters (vol. 1). In M. J. Kolen and W. Lee (Eds.), *Mixed-Format Tests: Psychometric Properties with a Primary Focus on Equating*, pp. 213–236. Iowa City, IA: CASMA, The University of Iowa.

Liu, C. and M. J. Kolen (2020). A new statistic for selecting the smoothing parameter for polynomial loglinear equating under the random groups design. *Journal of Educational Measurement 57*(3), 458–479.

Liu, J., T. Moses, and A. Low (2009). Evaluation of the effects of loglinear smoothing models on equating functions in the presence of structured data irregularities. *ETS Research Report 2009*(22), i–31.

Liu, Y. and A. Maydeu-Olivares (2012). Local dependence diagnostics in IRT modeling of binary data. *Measurement 73*(2), 254–274.

Livingston, S. (1992). Small-sample equating with log-linear smoothing. *ETS Research Report 1992*(1), i–10.

Livingston, S. (1993). An empirical tryout of kernel equating. *ETS Research Report 1993*(33), i–9.

Livingston, S. A. and N. J. Feryok (1987). Univariate vs. bivariate smoothing in frequency estimation equating. *ETS Research Report Series 1987*(2), i–21.

Lord, F. (1980). *Applications of Item Response Theory to Practical Testing Problems*. Hillsdale, NJ: Lawrence Erlbaum Associates.

Lord, F. and M. Novick (1968). *Statistical Theories of Mental Test Scores*. Reading, MA: Addison-Wesley.

Lord, F. and M. Wingersky (1984). Comparison of IRT true-score and equipercentile observed-score "equatings". *Applied Psychological Measurement 8*(4), 453–461.

Lord, F. M. (1950). Notes on comparable scales for test scores. Technical report, Educational Testing Service.

Lord, F. M. (1965). A strong true-score theory with applications. *Psychometrika 30*, 239–270.

Loyd, B. H. and H. Hoover (1980). Vertical equating using the Rasch model. *Journal of Educational Measurement 17*(3), 179–193.

Lu, R., S. Haberman, H. Guo, and J. Liu (2015). Use of jackknifing to evaluate effects of anchor item selection on equating with the nonequivalent groups with anchor test (NEAT) design. *ETS Research Report Series 2015*(1), 1–12.

Lukacs, E. (1970). *Characteristic Functions* (2nd ed.). A Charles Griffin Book. London: Hodder Arnold.

Lyrén, P.-E. and R. K. Hambleton (2011). Consequences of violated equating assumptions under the equivalent groups design. *International Journal of Testing 11*(4), 308–323.

Marco, G. L. (1977). Item characteristic curve solutions to three intractable testing problems. *Journal of Educational Measurement 14*(2), 139–160.

Marcq, K. and B. Andersson (2022). Standard errors of kernel equating: Accounting for bandwidth estimation. *Applied Psychological Measurement 46*(3), 200–218.

Masters, G. (1982). A Rasch model for partial credit scoring. *Psychometrika 47*, 149–174.

Maydeu-Olivares, A. (2013). Goodness-of-fit assessment of item response theory models. *Measurement: Interdisciplinary Research and Perspectives 11*(3), 71–101.

McCullagh, P. (2002). What is a statistical model? (with discussion). *The Annals of Statistics 30*, 1225–1310.

McLachlan, G. and D. Peel (2004). *Finite Mixture Models*. New York: John Wiley & Sons.

Michaelides, M. and E. Haertel (2004). Sampling of common items: An unrecognized source of error in test equating. *National Center for Research on Evaluation, Standards, and Student Testing CSE report 636*, 1–31.

Michaelides, M. and E. Haertel (2014). Selection of common items as an unrecognized source of variability in test equating: A bootstrap approximation assuming random sampling of common items. *Applied Measurement in Education 27*(1), 46–57.

Mislevy, R. J. and R. D. Bock (1990). *BILOG 3: Item analysis and test scoring with binary logistic models*. Scientific Software International.

Morris, G. (1982). On the foundations of test equating. In P. Holland and D. Rubin (Eds.), *Test Equating*, pp. 418–432. New York: Academic Press.

Moses, T. and P. Holland (2007). Kernel and traditional equipercentile equating with degrees of premoothing. *ETS Research Report 2007*(15), i–39.

Moses, T. and P. Holland (2009). Selection strategies for univariate loglinear smoothing models and their effect on equating function accuracy. *Journal of Educational Measurement 46*(2), 159–176.

Moses, T. and P. Holland (2010). A comparison of statistical selection strategies for univariate and bivariate log-linear models. *British Journal of Mathematical and Statistical Psychology 63*, 557–574.

Moses, T. and J. Liu (2011). Smoothing and equating methods applied to different types of test score distributions and evaluated with respect to multiple equating criteria. *ETS Research Report 2011*(20), i–25.

Moses, T. and W. Zhang (2011). Standard errors of equating differences: Prior developments, extensions, and simulations. *Journal of Educational and Behavioral Statistics 36*(6), 779–803.

Mosteller, F. and C. Youtz (1961, 12). Tables of the Freeman-Tukey transformations for the binomial and Poisson distributions*. *Biometrika 48*(3-4), 433–440.

Muraki, E. (1992). A generalized partial credit model: Application of an EM algorithm. *Applied Psychological Measurement 16*, 159–176.

Neammanee, K. (2005). A refinement of normal approximation to Poisson binomial. *International Journal of Mathematics and Mathematical Sciences 5*, 717–728.

Nedelman, J. and T. Wallenius (1986). Bernoulli trials, Poisson trials, surprising variances, and Jensen's inequality. *The American Statistician 40*, 286–289.

Nering, M. and R. Ostini (2010). *Handbook of Polytomous Item Response Theory Models*. New York: Routledge/Taylor & Francis Group.

Ogasawara, H. (2000). Asymptotic standard errors of IRT equating coefficients using moments. *Economic Review (Otaru University of Commerce) 51*(1), 1–23.

Ogasawara, H. (2003). Asymptotic standard errors of IRT observed-score equating methods. *Psychometrika 68*, 193–211.

Orlando, M. and D. Thissen (2000). Likelihood-based item-fit indices for dichotomous item response theory models. *Applied Psychological Measurement 24*(1), 50–64.

Penfield, R. D. (2014). An NCME instructional module on polytomous item response theory models. *Educational Measurement: Issues and Practice 33*(1), 36–48.

Petersen, N., M. Kolen, and H. Hoover (1989). Scaling, norming and equating. In R. L. Linn (Ed.), *Educational Measurement (3rd ed.)*, pp. 221–262. New York: MacMillan.

Qian, J., A. A. von Davier, and Y. Jiang (2013). Achieving a stable scale for an assessment with multiple forms: Weighting test samples in IRT linking. In R. Millsap, L. van der Ark, D. Bolt, and C. Woods (Eds.), *New Developments in Quantitative Psychology. Springer Proceedings in Mathematics and Statistics*, Volume 66, pp. 171–185. New York: Springer.

R Core Team (2024). *R: A Language and Environment for Statistical Computing*. Vienna, Austria: R Foundation for Statistical Computing.

Rajagopalan, B. and U. Lall (1995). A kernel estimator for discrete distributions. *Journal of Nonparametric Statistics 4*(4), 409–426.

Rao, B. L. S. P. (1983). *Nonparametric Functional Estimation.* Probability and Mathematical Statistics. Orlando, FL: Academic Press.

Rao, C. R. (1973). *Linear Statistical Inference and Applications.* New York: Wiley.

Revelle, W. (2023). psych: Procedures for pscyhological, psychometric, and personality research. https://CRAN.R-project.org/package=psych. R package version 2.3.3.

Revelle, W. and T. Rocklin (1979). Very simple structure: An alternative procedure for estimating the optimal number of interpretable factors. *Multivariate Behavioral Research 14*(4), 403–414.

Rizopoulos, D. (2006). ltm: An R package for latent variable modeling and item response theory analyses. *Journal of Statistical Software 17*(5), 1–25.

Rosenbaum, P. and D. Thayer (1987). Smoothing the joint and marginal distributions of scored two-way contingency tables in test equating. *British Journal of Mathematical and Statistical Psychology 40*, 43–49.

Rosenbaum, P. R. and D. B. Rubin (1983). The central role of the propensity score in observational studies for causal effects. *Biometrika 70*, 41–55.

Rosenblatt, M. (1956). Remarks on some nonparametric estimates of a density function. *The Annals of Mathematical Statistics 27*(3), 832–837.

Rupp, A. A. and B. D. Zumbo (2006). Understanding parameter invariance in unidimensional IRT models. *Educational and Psychological Measurement 66*(1), 63–84.

Samejima, F. (1969). Estimation of latent ability using a response pattern of graded scores. *Psychometrika monograph supplement 34*, 1–97.

Samuels, S. M. (1965). On the number of successes in independent trials. *The Annals of Mathematical Statistics 36*, 1272–1278.

San Martín, E. and J. González (2022). A critical view on the NEAT equating design: Statistical modelling and identifiability problems. *Journal of Educational and Behavioral Statistics 47*(4), 406–437.

Sansivieri, V. and M. Wiberg (2017). IRT observed-score equating with the nonequivalent groups with covariates design. In L. A. van der Ark, M. Wiberg, S. Culpepper, J. Douglas, and W.-C. Wang (Eds.), *Quantitative Psychology. The 81st Annual Meeting of the Psychometric Society, Asheville, North Carolina, 2016*, pp. 275–286. Springer.

Sansivieri, V. and M. Wiberg (2019). Linking scales in item response theory with covariates. *Journal of Research in Education, Science and Technology 3*(2), 12–32.

Schwarz, G. (1978). Estimating the dimension of a model. *The Annals of Statistics 6*, 461–464.

Scott, D. (1992). *Multivariate Density Estimation*. New York, NY: John Wiley.

Scott, D. W. (2015). *Multivariate Density Estimation: Theory, Practice, and Visualization* (2 ed.). Wiley Series in Probability and Statistics. John Wiley & Sons: New York, NY.

Silverman, B. (1986). *Density Estimation for Statistics and Data Analysis*, Volume 3. London: Chapman and Hall.

Sinharay, S., S. Haberman, P. Holland, and C. Lewis (2012). A note on the choice of an anchor test in equating. *ETS Research Report Series 2012*(2), i–9.

Sinharay, S., P. W. Holland, and A. A. von Davier (2011). Evaluating the missing data assumptions of the chain and poststratification equating methods. In A. von Davier (Ed.), *Statistical Models for Test Equating, Scaling, and Linking*, pp. 281–296. New York: Springer.

Skaggs, G. and R. Lissitz (1986). IRT test equating: Relevant issues and a review of recent research. *Review of Educational Research 56*(4), 495.

Smith, A. E. and D. W. Coit (1997). Penalty functions. In T. Bäck, D. Fogel, and Z. Michalewicz (Eds.), *Handbook of Evolutionary Computation*, pp. C5.2:1–6. Boca Raton, FL: Taylor and Francis.

Steele, J. M. (1994). Le Cam's inequality and Poisson approximations. *The American Mathematical Monthly 101*, 48–54.

Stocking, M. and F. Lord (1983). Developing a common metric in item response theory. *Applied Psychological Measurement 7*(2), 201–210.

Stone, C. A. and B. Zhang (2003). Assessing goodness of fit of item response theory models: A comparison of traditional and alternative procedures. *Journal of Educational Measurement 40*(4), 331–352.

Stone, M. (1974). Cross-validatory choice and assessment of statistical predictions (with discussion). *Journal of the Royal Statistical Society Series B 1974*(36), 111–147.

Sundt, B. and R. Vernic (2009). *Recursions for Convolutions and Compound Distributions with Insurance Applications*. Springer: Berlin Heidelberg.

Terrell, G. R. and D. W. Scott (1992). Variable kernel density estimation. *The Annals of Statistics 20*(3), 1236–1265.

Thissen, D., M. Pommerich, K. Billeaud, and V. S. L. Williams (1995). Item response theory for scores on tests including polytomous items with ordered responses. *Applied Psychological Measurement 19*(1), 39–49.

Thomas, M. and A. Taub (1982). Calculating binomial probabilities when the trial probabilities are unequal. *Journal of Statistical Computation and Simulation 14*, 125–131.

Tong, Y. and M. Kolen (2005). Assessing equating results on different equating criteria. *Applied Psychological Measurement 29*(6), 418–432.

Tuerlinckx, F., F. Rijmen, G. Molenberghs, G. Verbeke, D. Briggs, W. van den Noortgate, M. Meulders, and P. De Boeck (2004). Estimation and software. In P. D. Boeck and M. Wilson (Eds.), *Explanatory Item Response Models: A Generalized Linear and Nonlinear Approach*, Volume 1, pp. 343–373. New York: Springer.

van der Linden, W. J. (2000). A test-theoretic approach to observed-score equating. *Psychometrika 65*(4), 437–456.

van der Linden, W. J. (2006). Equating scores from adaptive to linear tests. *Applied Psychological Measurement 30*(6), 493–508.

van der Linden, W. J. (2011). Local observed-score equating. In A. von Davier (Ed.), *Statistical Models for Test Equating, Scaling, and Linking*, pp. 201–223. New York: Springer.

van der Linden, W. J. (2013). Some conceptual issues in observed-score equating. *Journal of Educational Measurement 50*(3), 249–285.

van der Linden, W. J. (2016). *Handbook of Item Response Theory*. Three volume set. Boca Raton, FL: Chapman and Hall/CRC.

van der Linden, W. J. (2019). Lord's equity theorem revisited. *Journal of Educational and Behavioral Statistics 44*(4), 415–430.

van der Linden, W. J. (2022). What is actually equated in "test equating"? A didactic note. *Journal of Educational and Behavioral Statistics 47*(3), 353–362.

Varas, I., J. González, and F. A. Quintana (2019). A new equating method through latent variables. In M. Wiberg, S. Culpepper, R. Janssen, J. González, and D. Molenaar (Eds.), *Quantitative Psychology. The 83rd Annual Meeting of the Psychometric Society, New York, NY, 2018*, pp. 343–353. Cham: Springer.

Varas, I., J. González, and F. A. Quintana (2020). A Bayesian nonparametric latent approach for score distributions in test equating. *Journal of Educational and Behavioral Statistics 45*(6), 639–666.

Velicer, W. F. (1976). Determining the number of components from the matrix of partial correlations. *Psychometrika 41*(3), 321–327.

Verhelst, N. (2001). Testing the unidimensionality assumption of the Rasch model. *Methods of Psychological Research Online 6*, 231–271.

Volkova, A. Y. (1996). A refinement of the central limit theorem for sums of independent random indicators. *Theory of Probability and Its Applications 40*, 791–794.

von Davier, A. (2011). *Statistical Models for Test Equating, Scaling, and Linking*. New York: Springer.

von Davier, A. A. (2013). Observed-score equating: An overview. *Psychometrika 78*(4), 605–623.

von Davier, A. A., P. Holland, and D. Thayer (2004). *The Kernel Method of Test Equating*. New York: Springer.

von Davier, A. A. and N. Kong (2005). A unified approach to linear equating for the nonequivalent groups design. *Journal of Educational and Behavioral Statistics 30*(3), 313–342.

von Davier, A. A. and C. Wilson (2007). IRT true-score equating: A guide through assumptions and applications. *Educational and Psychological Measurement 67*(6), 940–957.

von Davier, M. and A. A. von Davier (2007). A unified approach to IRT scale linking and scale transformations. *Methodology: European Journal of Research Methods for the Behavioral and Social Sciences 3*(3), 115–124.

Wallin, G., J. Häggström, and M. Wiberg (2018). How to select the bandwidth in kernel equating? - An evaluation of five different methods. In M. Wiberg, S. Culpepper, R. Janssen, J. González, and D. Molenaar (Eds.), *Quantitative Psychology. The 82nd Annual Meeting of the Psychometric Society, Zurich, Switzerland, 2017*, pp. 91–100. Springer.

Wallin, G., J. Häggström, and M. Wiberg (2021). How important is the choice of bandwidth in kernel equating? *Applied Psychological Measurement 45*(7–8), 518–535.

Wallin, G. and M. Wiberg (2017). Nonequivalent groups with covariates design using propensity scores for kernel equating. In L. A. van der Ark, M. Wiberg, S. Culpepper, J. Douglas, and W.-C. Wang (Eds.), *Quantitative Psychology. The 81st Annual Meeting of the Psychometric Society, Asheville, North Carolina, 2016*, pp. 309–319. Cham: Springer.

Wallin, G. and M. Wiberg (2019). Kernel equating using propensity scores for non-equivalent groups. *Journal of Educational and Behavioral Statistics 44*(4), 390–414.

Wallin, G. and M. Wiberg (2023). Model misspecification and robustness of observed-score test equating using propensity scores. *Journal of Educational and Behavioral Statistics 48*(5), 603–635.

Wallin, G. and M. Wiberg (2024). Smoothing bivariate test score distributions - model selection targeting test score equating. *Journal of Educational and Behavioral Statistics*. In press.

Wallmark, J., M. Josefsson, and M. Wiberg (2023a). Efficiency analysis of item response theory kernel equating for mixed-format tests. *Applied Psychological Measurement 47*(7–8), 496–512.

Wallmark, J., M. Josefsson, and M. Wiberg (2023b). Kernel equating presmoothing methods: An empirical study with mixed-format test forms. In M. Wiberg, D. Molenaar, J. González, J.-S. Kim, and H. Hwang (Eds.), *Quantitative Psychology. The 87rd Annual Meeting of the Psychometric Society, Bologna, Italy, 2022*, pp. 49–59. Cham: Springer.

Walsh, J. E. (1955). Approximate probability values for observed number of "successes" from statistically independent binomial events with unequal probabilities. *Sankhya: The Indian Journal of Statistics 15*, 281–290.

Wand, M. P. and M. C. Jones (1995). *Kernel Smoothing* (1st ed ed.). Monographs on Statistics and Applied Probability 60. Springer US.

Wang, T. (2011). An alternative continuization method: The continuized loglinear method. In A. von Davier (Ed.), *Statistical Models for Test Equating, Scaling, and Linking*, pp. 141–158. New York: Springer.

Wang, Y. (1993). On the number of successes in independent trials. *Statistica Sinica 3*, 295–312.

Wansouwé, W. E., S. M. Somé, and C. C. Kokonendji (2022). *Ake: Associated kernel estimators*. R package version 1.0.1.

Wiberg, M. (2012). Can a multidimensional test be evaluated with unidimensional item response theory? *Educational Research and Evaluation 18*(4), 307–320.

Wiberg, M. (2016). Alternative linear item response theory observed-score equating methods. *Applied Psychological Measurement 40*(3), 180–199.

Wiberg, M. (2017). Monitoring the equating transformation to ensure quality over time. *Paper presented at National Council of Measurement in Education San Antonio, Texas, April 26–30, 2017*.

Wiberg, M. and K. Bränberg (2015). Kernel equating under the non-equivalent groups with covariates design. *Applied Psychological Measurement 39*(5), 349–361.

Wiberg, M. and J. González (2016). Statistical assessment of estimated transformations in observed-score equating. *Journal of Educational Measurement 53*(1), 106–125.

Wiberg, M. and J. González (2021). Possible factors which may impact kernel equating of mixed-format tests. In M. Wiberg, D. Molenaar, J. González, U. Böckenholt, and J. Kim (Eds.), *Quantitative Psychology. The 85th Annual Meeting of the Psychometric Society, Virtual, 2020*, pp. 199–206. Cham: Springer.

Wiberg, M. and W. J. van der Linden (2011). Local linear observed-score equating. *Journal of Educational Measurement 48*, 229–254.

Wiberg, M., W. J. van der Linden, and A. A. von Davier (2014). Local observed-score kernel equating. *Journal of Educational Measurement 51*, 57–74.

Wiberg, M. and A. A. von Davier (2017). Examining the impact of covariates on anchor tests to ascertain quality over time in a college admissions test. *International Journal of Testing 17*, 105–126.

Wilk, M. and R. Gnanadesikan (1968). Probability plotting methods for the analysis of data. *Biometrika 55*(1), 1–17.

Wright, B. D. and G. N. Masters (1982). *Rating Scale Analysis*. Chicago, IL: MESA press.

Yen, W. M. (1984a). Effects of local item dependence on the fit and equating performance of the three-parameter logistic model. *Applied Psychological Measurement 8*(2), 125–145.

Yen, W. M. (1984b). Obtaining maximum likelihood trait estimates from number-correct scores for the three-parameter logistic model. *Journal of Educational Measurement 21*, 93–111.

Index

Adaptive kernel, 81, 82, 152, 153, 155, 156, 158
Akaike Information Criterion (AIC), 46, 171, 180
Anchor test, 12, 164
Average mean absolute difference (AMAD), 134
Average mean signed difference (AMSD), 134
Average point absolute difference, 134
Average root mean squared difference (ARMSD), 134

Bandwidth selection, 26, 87, 88, 150, 188
Bayesian Information Criterion (BIC), 171, 180
Bayesian information criterion (BIC), 46
Beta4 model, 41, 48, 144, 147, 158
Bias, 28, 93, 128, 135, 136
Bootstrap, 124
Bootstrap SEE, 158, 199

Chained equating (CE), 59, 105, 121, 168, 192, 195
Continuization, 8, 26, 77, 150, 188
Counterbalanced (CB) design, 11, 59
Cross-validation (CV), 91, 192
Cumulants, 84, 130

Data collection design, 9
Delta method, 116
Design function (DF), 25, 56–60, 62, 64, 117
Difference that matters (DTM), 114, 136

Difficulty parameter, 43
Discrete kernel estimators, 39, 47, 145, 148
Discrimination parameter, 43, 45, 108
Double smoothing, 93, 192

EDIFF, 133
Epanechnikov kernel, 81, 152, 153, 155, 156, 158
Equating indices, 136, 203
Equating transformation, 7, 13, 27, 188
Equating-specific measures, 113, 132, 198
Equipercentile equating, 14
Equity, 5, 114
Equivalent groups (EG) design, 10, 25, 57, 119, 141
Estimated probabilities, 64, 66
Estimating score probabilities, 25, 147, 148, 185, 186
Evaluating the equating transformation, 15, 27, 156, 198

Freeman-Tukey residuals, 46, 160

Gaussian kernel, 79, 82, 192

Haebara method, 109

IRT, 7, 42–44, 48, 66, 107, 168, 173, 175, 181, 182, 186, 194
Item characteristic curve (ICC), 44

Likelihood cross-validation, 92
Linear equating transformation, 13, 103

Linear interpolation, 8, 14, 26
Linking, 108, 109
Local equating, 110
Log-linear model, 8, 24, 26, 36, 37,
46, 168, 169, 171, 185, 190
Logistic kernel, 79, 192
Lord-Wingersky algorithm, 66, 70

Mean absolute difference (MAD),
126, 203
Mean signed difference (MSD), 126,
203
Mean squared error (MSE), 28, 93,
128, 129, 135, 136
Mean-mean method, 108, 194
Mean-sigma method, 108, 194
Minimizing a penalty function, 90,
150, 192
Moments, 108, 114

Non-equivalent groups with anchor
test (NEAT) design, 12, 25,
37, 59, 60, 64, 163, 165, 168
Non-equivalent groups with
covariates (NEC) design,
12, 25, 62, 64
Non-equivalent groups with
covariates (NEC) dsign, 121

Percent relative error (PRE), 28,
114, 136, 156, 198
Percentile rank method (PRM), 83
Poststratification equating (PSE),
59, 168, 190, 194
Presmoothing, 23, 33, 35, 52, 143,
168

Root mean squared difference
(RMSD), 126, 203
Root mean squared error (RMSE),
28, 129, 135, 136
Rule-based bandwidth selection, 94,
95, 192

Score scale, 7
Single group (SG) design, 10, 25, 58
Standard error (SE), 28, 129, 130,
135, 136
Standard error of equating (SEE),
115, 122, 136, 199, 200
Standard error of equating
differences (SEED), 28, 126,
201
Statistical measures, 127, 135, 198
Stocking and Lord method, 109, 194
Symmetric equating, 7

Uniform kernel, 80, 192